Bat Conservation

Bat Conservation

Global evidence for the effects of interventions

Anna Berthinussen, Olivia C. Richardson
and John D. Altringham

Synopses of Conservation Evidence, Volume 5

Pelagic Publishing | www.pelagicpublishing.com

Published by Pelagic Publishing
www.pelagicpublishing.com
PO Box 725, Exeter EX1 9QU

Bat Conservation
Global evidence for the effects of interventions
Synopses of Conservation Evidence, Volume 5
www.conservationevidence.com

ISBN 978-1-907807-89-3 (Pbk)
ISBN 978-1-907807-90-9 (Hbk)
ISBN 978-1-907807-91-6 (ePub)
ISBN 978-1-907807-92-3 (Kindle)

Series Editor: William J. Sutherland

Copyright © 2014 William J. Sutherland

This book should be quoted as Berthinussen, A., Richardson, Olivia C. and Altringham, John D. (2014) *Bat Conservation: Global evidence for the effects of interventions*. Exeter: Pelagic Publishing.

All rights reserved. No part of this document may be produced, stored in a retrieval system, or transmitted in any form or by any means, electronic, mechanical, photocopying, recording or otherwise without prior permission from the publisher. While every effort has been made in the preparation of this book to ensure the accuracy of the information presented, the information contained in this book is sold without warranty, either express or implied. Neither the author, nor Pelagic Publishing, its agents and distributors will be held liable for any damage or loss caused or alleged to be caused directly or indirectly by this book.

British Library Cataloguing in Publication Data
A catalogue record for this book is available from the British Library.

Cover image: *Megaderma spasma*, the lesser false vampire bat, roosting in a tourist shelter in Baluran National Park, in Java, Indonesia. Photograph by Manuel Ruedi.

Contents

Advisory board ix
About the authors x
Acknowledgements xi
About this book xii

1 Threat: Residential and commercial development 1
    Key messages 1
    1.1 Conserve existing roosts within developments 2
    1.2 Retain or relocate access points to bat roosts 2
    1.3 Create alternative roosts within buildings 2
    1.4 Change timing of building works 2
    1.5 Conserve old buildings or structures as roosting sites for bats 3
    1.6 Maintain bridges and retain crevices for roosting 3
    1.7 Protect brownfield sites 3
    1.8 Provide foraging habitat in urban areas 4
    1.9 Retain or replace existing bat commuting routes 5

2 Threat: Agriculture 6
    Key messages – Land use change 6
    Key messages – Intensive farming 6

    Land use change 7
    2.1 Conserve old buildings or structures as roosting sites for bats 7
    2.2 Retain dead/old trees with hollows and cracks as roosting sites for bats 7
    2.3 Retain or plant trees to replace foraging habitat for bats 7
    2.4 Protect or create wetlands as foraging habitat for bats 9
    2.5 Retain or replace existing bat commuting routes 10

    Intensive farming 10
    2.6 Convert to organic farming 10
    2.7 Introduce agri-environment schemes 12
    2.8 Encourage agroforestry 13

3 Threat: Energy production – wind turbines 16
    Key messages 16
    3.1 Modify turbine design to reduce bat fatalities 17
    3.2 Modify turbine placement to reduce bat fatalities 17

      3.3 Leave a minimum distance between turbines and habitat features used by bats   18

      3.4 Deter bats from turbines using radar   18

      3.5 Deter bats from turbines using ultrasound   19

      3.6 Remove turbine lighting to avoid attracting bats   21

      3.7 Switch off turbines at low wind speeds to reduce bat fatalities   21

      3.8 Automatically switch off wind turbines when bat activity is high   23

      3.9 Close off nacelles in wind turbines to prevent roosting bats   23

4 Threat: Energy production – mining   24

      Key messages   24

      4.1 Legally protect bat hibernation sites in mines from reclamation   24

      4.2 Provide artificial hibernacula to replace roosts lost in reclaimed mines   24

      4.3 Relocate bats from reclaimed mines to new hibernation sites   24

5 Threat: Transportation and service corridors   25

      Key messages – Roads   25

      5.1 Install underpasses as road crossing structures for bats   26

      5.2 Install overpasses as road crossing structures for bats   28

      5.3 Install bat gantries or bat bridges as road crossing structures for bats   28

      5.4 Install green bridges as road crossing structures for bats   29

      5.5 Install hop-overs as road crossing structures for bats   30

      5.6 Divert bats to safe crossing points with plantings or fencing   30

      5.7 Deter bats with lighting   31

      5.8 Replace or improve habitat for bats around roads   31

6 Threat: Biological resource use   32

      Key messages – Hunting   32

      Key messages – Guano harvesting   32

      Key messages – Logging and wood harvesting   32

      Hunting   33

      6.1 Introduce and enforce legislation to control hunting of bats   34

      6.2 Educate local communities about bats and hunting   34

      6.3 Introduce sustainable harvesting of bats   34

      Guano harvesting   34

      6.4 Introduce and enforce legislation to regulate harvesting of bat guano   35

      6.5 Introduce sustainable harvesting of bat guano   35

      Logging and wood harvesting   35

      6.6 Use selective harvesting/reduced impact logging instead of clearcutting   35

6.7 Use shelterwood cutting instead of clearcutting	39
6.8 Thin trees within forest	41
6.9 Manage forest or woodland edges for bats	43
6.10 Retain deadwood/snags within forests for roosting bats	43
6.11 Replant native trees	44
6.12 Retain residual tree patches in logged areas	44
6.13 Incorporate forested corridors or buffers into logged areas	45
7 Threat: Human disturbance – caving and tourism	**48**
Key messages	48
7.1 Use cave gates to restrict public access	48
7.2 Maintain microclimate at hibernation/roost sites	52
7.3 Impose restrictions on cave visits	52
7.4 Educate the public to reduce disturbance to hibernating bats	53
7.5 Legally protect bat hibernation sites	53
7.6 Provide artificial hibernacula for bats to replace disturbed sites	53
8 Threat: Natural system modification – natural fire and fire suppression	**54**
Key messages	54
8.1 Use prescribed burning	54
9 Threat: Invasive species and disease	**58**
Key messages – Invasive species	58
Key messages – White-nose syndrome	58
Invasive species	**58**
9.1 Remove invasive plant species	58
9.2 Control invasive predators	59
9.3 Translocate to predator or disease free areas	60
White-nose syndrome	**60**
9.4 Control anthropogenic spread	61
9.5 Increase population resistance	61
9.6 Cull infected bats	61
9.7 Modify cave environments to increase bat survival	61
10 Threat: Pollution	**63**
Key messages – Domestic and urban waste water	63
Key messages – Agricultural and forestry effluents	63
Key messages – Light and noise pollution	63
Key messages – Timber treatments	64
Domestic and urban waste water	**64**

	10.1 Change effluent treatments	64
	Agricultural and forestry effluents	65
	10.2 Introduce legislation to control use	65
	10.3 Change effluent treatments	65
	Light and noise pollution	65
	10.4 Leave bat roosts, roost entrances and commuting routes unlit	66
	10.5 Minimize light pollution	66
	10.6 Restrict timing of lighting	67
	10.7 Use low pressure sodium lamps or use UV filters	67
	10.8 Impose noise limits in proximity to roosts and bat habitats	67
	Timber treatments	67
	10.9 Use mammal-safe timber treatments in roof spaces	68
	10.10 Restrict timing of timber treatment application	69
11	Providing artificial roost structures for bats	70
	Key messages	70
	11.1 Provide artificial roost structures for bats	70
12	Education and awareness raising	79
	Key messages	79
	12.1 Provide training to professionals	79
	12.2 Educate homeowners about building and planning laws	79
	12.3 Educate to improve public perception and improve awareness	80
	Index	81

Advisory board

We thank the following people for advising on the scope and content of this synopsis.

Brock Fenton, University of Western Ontario, Canada
Karen Haysom and colleagues, Bat Conservation Trust, UK
David Jacobs, University of Cape Town, South Africa
Javier Juste, Evolutionary Biology Unit (CSIC), Spain
Gareth Jones, University of Bristol, UK
Kirsty Park, University of Stirling, UK
Stuart Parsons, University of Auckland, New Zealand
Paul Racey, University of Exeter, UK
Danilo Russo, Federico II University of Naples, Italy

About the authors

Anna Berthinussen is a postdoctoral researcher in the School of Biology, University of Leeds, UK.

Olivia C. Richardson is a former Masters student in the School of Biology, University of Leeds, UK.

John D. Altringham is Professor of Animal Ecology and Conservation in the School of Biology, University of Leeds, UK.

Acknowledgements

This synopsis was prepared with funding from Natural England. Additional funding was provided by Pettersson Elektronik (Uppsala Science Park, Dag Hammarskjolds v. 34A, S-751 83 UPPSALA, SWEDEN, http://www.batsound.com). The Conservation Evidence project has also received funding from the Natural Environment Research Council (NERC) and Arcadia.

We also thank Dr Lynn Dicks, Professor William Sutherland, Dr Stephanie Prior and Dr Rebecca Smith for their help and advice.

About this book

The purpose of Conservation Evidence synopses

Conservation Evidence synopses **do**	Conservation Evidence synopses **do not**
Bring together scientific evidence captured by the Conservation Evidence project (over 4,000 studies so far) on the effects of interventions to conserve wildlife	Include evidence on the basic ecology of species or habitats, or threats to them
List all realistic interventions for the species group or habitat in question, regardless of how much evidence for their effects is available	Make any attempt to weight or prioritize interventions according to their importance or the size of their effects
Describe each piece of evidence, including methods, as clearly as possible, allowing readers to assess the quality of evidence	Weight or numerically evaluate the evidence according to its quality
Work in partnership with conservation practitioners, policymakers and scientists to develop the list of interventions and ensure we have covered the most important literature	Provide recommendations for conservation problems, but instead provide scientific information to help with decision-making

Who this synopsis is for

If you are reading this, we hope you are someone who has to make decisions about how best to support or conserve biodiversity. You might be a land manager, a conservationist in the public or private sector, a farmer, a campaigner, an advisor or consultant, a policymaker, a researcher or someone taking action to protect your own local wildlife. Our synopses summarize scientific evidence relevant to your conservation objectives and the actions you could take to achieve them.

We do not aim to make your decisions for you, but to support your decision-making by telling you what evidence there is (or isn't) about the effects that your planned actions could have.

When decisions have to be made with particularly important consequences, we recommend carrying out a systematic review, as the latter is likely to be more comprehensive than the summary of evidence presented here. Guidance on how to carry out systematic reviews can be found from the Centre for Evidence-Based Conservation at the University of Bangor (www.cebc.bangor.ac.uk).

The Conservation Evidence project
The Conservation Evidence project has three parts:

- An online, **open access journal** *Conservation Evidence* publishes new pieces of research on the effects of conservation management interventions. All our papers are written by, or in conjunction with, those who carried out the conservation work and include some monitoring of its effects.
- An ever-expanding **database of summaries** of previously published scientific papers, reports, reviews or systematic reviews that document the effects of interventions.
- **Synopses** of the evidence captured in parts one and two on particular species groups or habitats. Synopses bring together the evidence for each possible intervention. They are freely available online and available to purchase in printed book form.

These resources currently comprise over 4,000 pieces of evidence, all available in a searchable database on the website www.conservationevidence.com.

Alongside this project, the Centre for Evidence-Based Conservation (www.cebc.bangor.ac.uk) and the Collaboration for Environmental Evidence (www.environmentalevidence.org) carry out and compile systematic reviews of evidence on the effectiveness of particular conservation interventions. These systematic reviews are included on the Conservation Evidence database.

Of the 78 bat conservation interventions identified in this synopsis, none are currently subject to systematic review.

Two interventions that we feel would benefit significantly from systematic reviews are the provision of bat boxes and the provision of artificial hibernacula, since both are widely practiced.

Scope of the Bat Conservation synopsis
This synopsis covers evidence for the effects of conservation interventions for native, wild bats.

It is restricted to evidence captured on the website www.conservationevidence.com. It includes papers published in the journal *Conservation Evidence*, evidence summarized on our database and systematic reviews collated by the Collaboration for Environmental Evidence.

Evidence from all around the world is included. If there appears to be a bias towards evidence from northern European or North American temperate environments, this reflects a current bias in the published research that is available to us.

How we decided which bat conservation interventions to include
Our list of interventions has been agreed in partnership with an Advisory Board made up of international conservationists and academics with expertise in bat conservation. Although the list of interventions may not be exhaustive, we have tried to include all actions that have been carried out or advised to support populations or communities of wild bats.

How we reviewed the literature
In addition to evidence already captured by the Conservation Evidence project, we have searched the following sources for evidence relating to bat conservation:

- Five specialist bat or mammal journals from their first publication date to the end of 2012 (*Acta Chiropterologica, Journal of Mammalogy, Mammalia, Mammal Review, Mammalian Biology*).
- Ten general ecology and conservation journals over the same time period.
- Where we knew of an intervention which we had not captured evidence for, we performed keyword searches on ISI Web of Science and www.scholar.google.com for this intervention.
- Reports published by organizations conducting practical and research work in bat conservation.
- In total, 101 individual studies are covered in this synopsis, all included in full or in summary on the Conservation Evidence website.

The criteria for inclusion of studies in the Conservation Evidence database are as follows:

- There must have been an intervention that conservationists would do.
- Its effects must have been monitored quantitatively.
- In some cases, where a body of literature has strong implications for conservation of a particular species group or habitat, although it does not directly test interventions for their effects, we refer the reader to this literature.

How the evidence is summarized

Conservation interventions are grouped primarily according to the relevant direct threats, as defined in the International Union for Conservation of Nature (IUCN)'s Unified Classification of Direct Threats (www.iucnredlist.org/technical-documents/classification-schemes). In most cases, it is clear which main threat a particular intervention is meant to alleviate or counteract.

Not all IUCN threat types are included, only those that threaten bats, and for which realistic conservation interventions have been suggested.

Some important interventions can be used in response to many different threats, and it would not make sense to split studies up depending on the specific threat they were studying. We have separated out two important categories of conservation action, as defined by the IUCN, which are relevant to a variety of situations, habitats and threats. They are: 'Providing artificial roost structures for bats' and 'Education and awareness raising'. These respectively match the following categories of conservation actions defined by the IUCN: 'Species management: species recovery' and 'Education and awareness'.

Normally, no intervention or piece of evidence is listed in more than one place, and when there is ambiguity about where a particular intervention should fall there is clear cross-referencing. Some studies describe the effects of multiple interventions. When this is the case, cross-referencing is again used to direct readers to the other interventions investigated. Where a study has not separated out the effects of different interventions, the study is included in the section on each intervention, but the fact that several interventions were used is highlighted.

In the text of each section, studies are presented in chronological order, so the most recent evidence is presented at the end. The summary text at the start of each section groups studies according to their findings.

At the start of each chapter, a series of **key messages** provides a rapid overview of the evidence. These messages are condensed from the summary text for each intervention.

Background information is provided where we feel recent knowledge is required to interpret the evidence. This is presented separately and relevant references are included in the reference list at the end of each intervention section.

References containing evidence for the effects of interventions are summarized in more detail on the Conservation Evidence website. In electronic versions of the synopsis, they are hyperlinked directly to the Conservation Evidence summary. If you do not have access to the electronic version of the synopsis, typing the first author's name into the 'Quick Search' facility on www.conservationevidence.com is the quickest way to locate summaries.

The information in this synopsis is available in three ways:

- As a book, printed by Pelagic Publishing and for sale from www.pelagicpublishing.com.
- As a PDF to download from www.conservationevidence.com.
- As text for individual interventions on the searchable database at www.conservationevidence.com.

Terminology used to describe evidence

Unlike systematic reviews of particular conservation questions, we do not quantitatively assess the evidence, or weight it according to quality. However, to allow you to interpret evidence, we make the size and design of each trial we report clear. The table below defines the terms that we have used to do this.

The strongest evidence comes from randomized, replicated, controlled trials with paired sites and before-and-after monitoring.

Term	Meaning
Site comparison	A study that considers the effects of interventions by comparing sites that have historically had different interventions or levels of intervention.
Replicated	The intervention was repeated on more than one individual or site. In conservation and ecology, the number of replicates is much smaller than it would be for medical trials (when thousands of individuals are often tested). If the replicates are sites, pragmatism dictates that between five and ten replicates is a reasonable amount of replication, although more would be preferable. We provide the number of replicates wherever possible, and describe a replicated trial as 'small' if the number of replicates is small relative to similar studies of its kind.
Controlled	Individuals or sites treated with the intervention are compared with control individuals or sites not treated with the intervention.

Paired sites	Sites are considered in pairs, within which one was treated with the intervention and the other was not. Pairs of sites are selected with similar environmental conditions, such as soil type or surrounding landscape. This approach aims to reduce environmental variation and make it easier to detect a true effect of the intervention.
Randomized	The intervention was allocated randomly to individuals or sites. This means that the initial condition of those given the intervention is less likely to bias the outcome.
Before-and-after trial	Monitoring of effects was carried out before and after the intervention was imposed.
Review	A conventional review of literature. Generally, these have not used an agreed search protocol or quantitative assessments of the evidence.
Systematic review	A systematic review follows an agreed set of methods for identifying studies and carrying out a formal 'meta-analysis'. It will weight or evaluate studies according to the strength of evidence they offer, based on the size of each study and the rigour of its design. All environmental systematic reviews are available at: www.environmentalevidence.org
Study	If none of the above apply, for example a study looking at the number of people that were engaged in an awareness raising project.

Taxonomy
We do not update taxonomy, but employ species names used in the original paper. Where possible, common names and Latin names are both given the first time each species is mentioned within each intervention.

Significant results
Throughout the synopsis we have quoted results from papers. Unless specifically stated, these results reflect statistical tests performed on the results.

Interpretation of evidence
Care must be taken when interpreting some of the evidence provided. Studies do not always measure the most appropriate metric or assess at the population level. For example, a small proportion of bats using a bridge to cross a road is not an effective intervention if a greater proportion are being killed by traffic on the road below, with a negative overall impact on local bat populations.

IMPORTANT NOTE – defining the phrase 'we found no evidence'
For some interventions we were unable to present any evidence for their effectiveness. This was because either no research had been done in these areas, or previous work did not meet the criteria for this synopsis, in that interventions were not tested directly and quantitatively, or results may not have been reported or made publicly available. This does not mean that these interventions may not be effective in bat conservation, or that such measures should be

abandoned; it simply highlights the need for robust monitoring in these areas to ensure that future conservation efforts will be appropriate and effective.

How you can help to change conservation practice

If you know of evidence relating to bat conservation that is not included in this synopsis, we invite you to contact us, via the www.conservationevidence.com website.

Following guidelines provided on the site, you can submit a summary of a previously published study, or submit a paper describing new evidence to the *Conservation Evidence* journal. We particularly welcome summaries written by the authors of papers published elsewhere, and papers submitted by conservation practitioners.

1 Threat: Residential and commercial development

Key messages

Conserve existing roosts within developments
We found no evidence for the effects of interventions to conserve existing roosts within developments.

Retain or relocate access points to bat roosts
We found no evidence for the effects of interventions to retain or relocate access points to bat roosts within developments.

Create alternative roosts within buildings
We found no evidence for the effects of interventions to provide alternative roosts within buildings.

Change timing of building works
We found no evidence for the effects of changing the timing of building works on bats.

Conserve old buildings or structures as roosting sites for bats
We found no evidence for the effects of conserving old buildings or structures as roosting sites for bats.

Maintain bridges and retain crevices for roosting
We found no evidence for the effects of maintaining old bridges and retaining crevices for roosting bats.

Protect brownfield sites
One study in the USA found bat activity within an urban wildlife refuge on an abandoned manufacturing site to be consistent with predictions across North America based on the availability of potential roosts.

Provide foraging habitat in urban areas
One site comparison study in the USA found higher bat activity in restored forest preserves in urban areas than in an unrestored forest preserve. One replicated, controlled, site comparison study in the UK found higher bat activity over green roofs in urban areas than conventional unvegetated roofs.

Retain or replace existing bat commuting routes
We found no evidence for the effects of retaining or replacing the original commuting routes of bats lost to residential or commercial development.

For all evidence relating to the use of **bat boxes/houses**, see 'Providing artificial roost structures for bats'.

1.1 Conserve existing roosts within developments

- We found no evidence for the effects of interventions to conserve existing roosts within residential or commercial developments.

Background
Many bat species are known to roost in the crevices and roof voids of buildings. Existing roosts can be conserved during residential or commercial developments, for example by retaining a roof space used as a roost during renovations. It is possible that with laws protecting bats in many countries, and with stricter planning and licensing requirements in the UK, for example, the consideration of bat roosts during residential and commercial development of buildings is benefitting bats. However, we found no studies examining the effects of attempts to conserve existing bat roosts in buildings during developments and encourage those working in this and related areas to publish their findings.

1.2 Retain or relocate access points to bat roosts

- We found no evidence for the effects of interventions to either retain or relocate access points to bat roosts within developments.

1.3 Create alternative roosts within buildings

- We found no evidence for the effects of interventions to provide alternative roosts within buildings.

Background
Alternative bat roosts may be created in buildings by providing suitable crevices or spaces. Bat bricks have been suggested as a way to incorporate bat roosts into newly built buildings, and are available for purchase. They are bricks with small crevices that bats may roost in or use as access to a roost in a cavity behind the brick. Heated bat boxes, which simulate the environmental conditions found in roosts within buildings, may also be used to replace roosts lost in developments. We found no studies examining the effects of interventions, including bat bricks or heated bat boxes, to provide alternative roosts within buildings.

1.4 Change timing of building works

- We found no evidence for the effects of changing the timing of building works on bats.

Background
To reduce disturbance to bats, building works may be avoided at times of the year when they are most vulnerable, such as during hibernation and the breeding season.

3 Threat: Residential and commercial development

1.5 Conserve old buildings or structures as roosting sites for bats

- We found no evidence for the effects of conserving old buildings or structures as roosting sites for bats.

1.6 Maintain bridges and retain crevices for roosting

- We found no evidence for the effects of interventions to maintain old bridges used as roosts by bats, or to retain crevices within them used as access points.

1.7 Protect brownfield sites

- One study in Denver, USA[1] found that the number and evenness of bat species within an urban wildlife refuge on an abandoned manufacturing site was consistent with predictions across North America based on the availability of potential roosts.

Background
'Brownfield sites' are previous industrial or commercial sites that have been abandoned and are available for reuse. These sites may be targeted for redevelopment in urban areas. They can support a high diversity of wildlife making them important sites for biodiversity and conservation. High insect numbers can provide important foraging habitat for bats, and derelict buildings may provide roosting opportunities.

A study in 1997 and 1998 in an urban wildlife refuge on the grounds of a former weapons manufacturing facility near Denver in the USA (1) found that the number and evenness of bat species was consistent with predictions across North America based on the availability of potential roosts. Details of the predictions are not given but the authors state that as expected from the availability of roosts (in a few dead trees and an abundance of buildings), three tree-roosting species and two species known to roost in buildings were captured or recorded, with big brown bats *Eptesicus fuscus* (common in urban areas) making up 86% of the captures. In total, 176 bats were captured and 955 bat passes were recorded. Activity was more than five times greater in areas of the refuge with tree or water habitat edges than in open prairie. Big brown bats commuted further from roosts in buildings within surrounding urban areas to the refuge (9–19 km) than typically reported for the species elsewhere (1–2 km). Bats were captured over two years from May to August on 53 nights at 18 sites. Mist nets were set up over water or by trees and shrubs within the refuge. Twelve big brown bats were captured and radio-tagged in 1998. Echolocation activity was recorded using bat detectors at eight sites with different habitat types within the refuge. Each site was sampled for a total of 90 minutes on 3–4 nights between June and August 1997. The chemical weapons facility was active up until 1985, and the site was designated and protected as a wildlife refuge with the passing of an Act in 1992. The refuge covers 6,900 ha consisting of grassland with scattered woodland and wetlands. It borders an urban area with a population of two million people, as well as industrial and agricultural land.

(1) Everette A.L., O'Shea T.J., Ellison L.E., Stone, L.A. & McCance J.L. (2001) Bat use of a high plains urban wildlife refuge. *Wildlife Society Bulletin*, 29, 967–973.

1.8 Provide foraging habitat in urban areas

- One site comparison study in the USA[1] found higher bat activity in restored forest preserves in urban areas than in an unrestored forest preserve. Different species responded differently to the changes in forest structure.

- One replicated, controlled, site comparison study in the UK[2] found significantly higher bat activity and more bat feeding events over green roofs in urban areas than conventional unvegetated roofs.

Background

Providing foraging habitat for bats in urban areas may reduce the impact of residential and commercial development. Existing foraging sites may be protected, or be replaced with suitable alternatives such as parks, woodland and wetlands. Bat activity was found to be higher in large parks in Mexico City than in natural forest or other urban habitats, although the number of species was higher in natural forest (Avila-Flores & Fenton 2005). Historic landscape parks in England, UK were found to contain higher proportions of rarer species than expected within the British bat fauna, with wetlands, woodland and linear elements being important (Glendell & Vaughan 2002). We found no studies examining the effects of creating parks, woodland or wetlands for bats in urban areas.

Avila-Flores R. & Fenton M.B. (2005) Use of spatial features by foraging insectivorous bats in a large urban landscape. *Journal of Mammalogy*, 86, 1193–1204.
Glendell M. & Vaughan N. (2002) Foraging activity of bats in historic landscape parks in relation to habitat composition and park management. *Animal Conservation*, 5, 309–316.

In a site comparison study in 2004–2005 in nine forest preserves within the Chicago metropolitan area, USA (1) the highest bat activity was recorded in two preserves that had undergone restoration with multiple prescribed burns, invasive plant species removal and snag recruitment (average 19 and 16 bat passes/preserve in 2004, average 7 and 18 bat passes/preserve in 2005). The lowest bat activity was recorded in a control site with no restoration (both years average one bat pass in total). Overall bat activity at all sites was positively related to prescribed burning, invasive species removal and small tree density (8–20 cm diameter at breast height) and negatively related to shrub density and clutter at heights of 0–6 m above the ground. Responses to woodland restoration varied among bat species. The eastern red bat *Lasiurus borealis* was positively associated with small and medium (20–33 cm) tree densities and negatively related to clutter at 0–9 m. *Myotis* spp. were positively associated with canopy cover, clutter at 6–9 m and small and medium tree densities. The silver-haired bat *Lasionycteris noctivagans* was positively associated with more open forests. The activity of the big brown bat *Eptesicus fuscus* was not associated with any vegetation variables in the study. The nine forest preserves varied in size from 10 to 260 ha. Fire suppression over the last 100 years had altered the structural diversity of the forests. Eight of the forest preserves were under management to restore forest to pre-European settlement conditions. Restoration practices were used to open canopy cover, reduce tree density and remove invasive plant species. Bats were monitored for four hours from sunset with bat detectors in June–September 2004 and May–August 2005 for five nights per site per year. Twenty randomly located 30 m line transects were sampled per site with four detectors placed 10 m apart along each transect.

5 Threat: Residential and commercial development

A replicated, controlled, site comparison study from May to September 2010 in urban areas of Greater London, UK (2) found significantly higher bat activity over 'biodiverse' green roofs planted with a variety of wild flowers, herbs, sedums, mosses and grasses than over conventional unvegetated roofs. An average of 8 bat call sequences and 0.56 feeding events/night were recorded over 'biodiverse' green roofs and 5 bat call sequences and 0.38 feeding events/night over conventional roofs. This was significant when a small amount (< 33%) of suitable bat habitat was located within 100 m of the roof. Bat activity and feeding events over 'sedum' green roofs planted with low-growing succulent plants (average 2 bat call sequences and 0.03 feeding events/night) did not differ from conventional roofs. Common pipistrelle *Pipistrellus pipistrellus* was most frequently recorded followed by soprano pipistrelle *Pipistrellus pygmaeus*, *Nyctalus/Eptesicus* spp. and Nathusius' pipistrelle *Pipistrellus nathusii*. All species were recorded feeding over biodiverse green roofs, but only common pipistrelles (and one record of a noctule bat *Nyctalus noctula*) were recorded over conventional and sedum green roofs. Roof height was found to negatively affect bat activity, with only one feeding event recorded over buildings more than two storeys high. Bat activity was recorded over 13 biodiverse, 9 sedum and 17 conventional roofs for 7 full nights. Conventional roofs were flat or shallow pitched with bitumen felt or paving slabs. All green roofs were low maintenance 'extensive' roofs, with shallow substrate (20–200 mm, but usually 35–75 mm).

(1) Smith D.A. & Gehrt S.D. (2010) Bat response to woodland restoration within urban forest fragments. *Restoration Ecology*, 18, 914–923.
(2) Pearce H. & Walters C. (2012) Do green roofs provide habitat for bats in urban areas? *Acta Chiropterologica*, 14, 469–478.

1.9 Retain or replace existing bat commuting routes

- We found no evidence for the effects of retaining or replacing the original commuting routes of bats lost to residential or commercial development.

Background
Linear features such as hedgerows and treelines provide important commuting routes for bats (Limpens & Kapteyn 1991, Verboom & Huitema 1997). Where original commuting routes cannot be retained, the use of artificial structures to replace them has been suggested. We found no evidence for either retaining original bat commuting routes or replacing them. For evidence relating to diverting bats using artificial structures as commuting routes see 'Threat: Transportation – Roads'.

Limpens H.J. & Kapteyn K. (1991) Bats, their behaviour and linear landscape elements. *Myotis*, 29, 39–48.
Verboom B. & Huitema H. (1997) The importance of linear landscape elements for the pipistrelle *Pipistrellus pipistrellus* and the serotine bat *Eptesicus serotinus*. *Landscape Ecology*, 12, 117–125.

2 Threat: Agriculture

Key messages – Land use change

Conserve old buildings or structures as roosting sites for bats
We found no evidence for the effects of conserving old buildings or structures on agricultural land as roosting sites for bats.

Retain old or dead trees with hollows and cracks as roosting sites for bats
We found no evidence for the effects of retaining old or dead trees on agricultural land as roosting sites for bats.

Retain or plant trees to replace foraging habitat for bats
We found no evidence for the effects of retaining trees as foraging habitat for bats. Two site comparison studies (one replicated) in Australia found no difference in bat activity and the number of bat species in agricultural areas revegetated with native plantings and over grazing land without trees. In both studies, bat activity was lower in plantings than in original forest and woodland remnants.

Protect or create wetlands as foraging habitat for bats
We found no evidence for the effects of protecting existing wetlands. One replicated, controlled, site comparison study in the USA found higher bat activity over heliponds and drainage ditches within a pine plantation than over natural wetlands. A replicated study in Germany found high levels of bat activity over constructed retention ponds compared to nearby vineyard sites, but comparisons were not made with natural pond sites.

Retain or replace existing bat commuting routes
We found no evidence for the effects of retaining or replacing the original commuting routes of bats lost due to agricultural land use change.

For all evidence relating to the use of **bat boxes/houses**, see 'Providing artificial roost structures for bats'.

Key messages – Intensive farming

Convert to organic farming
Four replicated, paired, site comparison studies on farms in the UK had inconsistent results. Two studies found higher bat abundance and activity on organic farms than conventional farms, and two studies showed no difference in bat abundance between organic and non-organic farms.

Introduce agri-environment schemes
One replicated, paired study in Scotland, UK found lower bat activity on farms participating in agri-environment schemes than on non-participating conventional farms.

7 Threat: Agriculture

Encourage agroforestry

Four replicated, site comparison studies (three in Mexico and one in Costa Rica) found no difference in bat diversity, the number of bat species and/or bat abundance between cacao, coffee or banana agroforestry plantations and native rainforest. One replicated, site comparison study in Mexico found higher bat diversity in native forest fragments than in coffee agroforestry plantations. One replicated, randomized, site comparison study in Costa Rica found lower bat diversity in native rainforest than in cacao agroforestry plantations. A replicated, site comparison study in Mexico found that bat diversity in coffee agroforestry plantations and native rainforest was affected by the proportion of each habitat type within the landscape. Three studies found that increasing management intensity on agroforestry plantations had a negative effect on some bat species, and a positive effect on others.

Land use change

2.1 Conserve old buildings or structures as roosting sites for bats

- We found no evidence for the effects of conserving old buildings or structures on agricultural land as roosting sites for bats.

> **Background**
> This intervention involves conserving roosting sites for bats in farm buildings, dry stone walls and natural structures found on farmland, such as sinkholes and caves.

2.2 Retain dead/old trees with hollows and cracks as roosting sites for bats

- We found no evidence for the effects of retaining old or dead trees with hollows and cracks as roosting sites for bats on agricultural land.

> See also 'Threat: Biological resource use – Logging and wood harvesting – Retain deadwood/snags within forests for roosting bats'.

2.3 Retain or plant trees to replace foraging habitat for bats

- We found no evidence for the effects of retaining trees as foraging habitat for bats.
- Two site comparison studies (one replicated) in Australia[1, 2] found no difference in bat activity and the number of bat species in agricultural areas revegetated with native plantings and over grazing land without trees. In both studies, bat activity was lower in plantings than in original forest and woodland remnants.

2.3 Retain or plant trees to replace foraging habitat for bats

Background
There is evidence that scattered trees and forest fragments are of conservation value and provide foraging habitat for bats in agricultural landscapes (Lumsden & Bennett 2005, Struebig *et al.* 2008, Fischer *et al.* 2010, Lentini *et al.* 2012). We found no intervention-based evidence for the effects of retaining trees or woodland patches within farms on bats. It has been suggested that planting trees in agricultural areas may replace lost foraging habitat. We found two studies that provide evidence for the effects on bats of creating plantations of native tree species in agricultural areas. For evidence relating to trees that are either retained or planted in agricultural systems to shade crops see 'Threat: Agriculture – Intensive Farming – Encourage agroforestry'.

Fischer J., Stott J. & Law B.S. (2010) The disproportionate value of scattered trees. *Biological Conservation*, 143, 1564–1567.
Lentini P.E., Gibbons P., Fischer J., Law B., Hanspach J. & Martin T.G. (2012) Bats in a farming landscape benefit from linear remnants and unimproved pastures. *PLoS ONE*, 7, e48201.
Lumsden L.F. & Bennett A.F. (2005) Scattered trees in rural landscapes: foraging habitat for insectivorous bats in south-eastern Australia. *Biological Conservation*, 122, 205–222.
Struebig M.J., Kingston T., Zubaid A., Mohd-Adnan A. & Rossiter S.J. (2008) Conservation value of forest fragments to Palaeotropical bats. *Biological Conservation*, 141, 2112–2126.

A replicated, site comparison study in late spring and early summer 2002 in an agricultural and forested area of New South Wales and Victoria, Australia (1), found that sites revegetated with native eucalypt plantings did not have significantly higher bat activity or more species than treeless grazed paddocks (average 87 vs. 50 bat passes/night and 5–7 vs. 5 species, respectively). Bat activity in revegetated sites was less than a third of that recorded in small remnants of original forest and woodland (average 302 bat passes/night). There was no significant difference between the number of bat species found in young plantings (average six species) and original remnants (average seven species). Old plantings had fewer bat species than remnant sites of a similar size. There was no significant difference in bat activity between large and small or young and old plantings. Twelve treatment classes including different sizes and ages of plantings, original remnants of forest and woodland, and grazed paddocks with and without trees were sampled with ten replicates per site. Bat activity was recorded at a single location at each site for one full night using bat detectors. The study area was dominated by grazing land with sparse remnants of forest and woodland vegetation. As part of a government initiative, extensive planting of locally indigenous tree species was carried out from the mid 1970s to 1991.

A site comparison study in summer 1999 in four agricultural sites revegetated with native bluegum *Eucalyptus globulus* plantations in Western Australia (2) found that bat activity was higher in plantations next to remnant vegetation than over agricultural grazing land, although differences were not tested for statistical significance (total 52 vs. 14 bat passes respectively). Bat activity was highest in remnant vegetation (75 bat passes), and lowest in plantations isolated from remnants of original vegetation and surrounded by grazing land (four bat passes). Similar numbers of species (two to four) were recorded in plantations and grazing land. Eight bat species were detected in original remnants, three of which were only found in this habitat type. The four study areas selected consisted of commercially established farm forestry plantations of four to six years in age, remnants of original native vegetation, and open grazing land. Within each study area, four treatment sites were sampled for one full night and bat activity recorded with bat detectors. The treatments sampled were remnant vegetation, plantations adjacent to remnant vegetation,

9 Threat: Agriculture

grazing land and plantations adjacent to grazing land. The results from all four study areas were pooled to give totals for each treatment.

(1) Law B.S. & Chidel M. (2006) Eucalypt plantings on farms: use by insectivorous bats in south-eastern Australia. *Biological Conservation*, 133, 236–249.
(2) Hobbs R., Catling P.C., Wombey J.C., Clayton M., Atkins L. & Reid A. (2003) Faunal use of bluegum (*Eucalyptus globulus*) plantations in southwestern Australia. *Agroforestry Systems*, 58, 195–212.

2.4 Protect or create wetlands as foraging habitat for bats

- One replicated, controlled, site comparison study in the USA[1] found higher bat activity over heliponds and drainage ditches within a pine plantation than over natural wetlands.

- One replicated study in Germany[2] found higher bat activity over constructed retention ponds than at sites in nearby vineyards. No comparisons were made between the artificial ponds and natural wetland sites.

In a replicated, controlled, site comparison study in June–July 2006 and 2007 in intensively managed pine plantations in North Carolina, USA (1) overall bat activity was found to be highest at modified water sources within the plantations than in natural wetlands. The total of averages for all species were 201 call sequences/site/night at heliponds, 61 call sequences/site/night at interior ditches, 60 call sequences/site/night at edge ditches, and 21 call sequences/site/night at natural wetlands. Modified water sources were either drainage ditches (1–2.5 m wide and 0.6–1.2 m deep) positioned every 80–100m within stands and along stand borders, or small ponds 'heliponds' (12 m x 24 m x 2.5 m deep) used by helicopters for the suppression of forest fires. The natural wetland site was a 350 ha remnant natural forested wetland adjacent to the plantation. Surveys were conducted on 116 nights over the two summers. On each night bat activity was sampled simultaneously from dusk until dawn with bat detectors at two of four water source types rotated in a random order (five heliponds, five interior ditches, five edge ditches and three natural wetlands). Bats were caught in mist nets at heliponds and natural wetland sites to confirm bat species presence. Seven species or species groups were identified. Insect abundance was measured using passive malaise and emergence traps, but did not differ across water source types.

A replicated study in summer 2009 at seven agricultural vineyard sites in Landau, Germany (2) found that total bat activity above artificial retention ponds was significantly higher than above nearby vineyard sites (average 1,543 vs. 25 s of recorded call sequences respectively). Ten species were recorded in total. Activity over retention ponds was 180 times higher than vineyard sites for *Pipistrellus* species, and 50 times higher for *Myotis* species. Foraging activity relative to the area of habitat available showed that retention ponds had on average a higher importance as bat foraging habitats than the complete vineyard. Seven retention ponds were sampled from the end of June to the end of August. All ponds had banks lined with trees or bushes, and ranged in size from 0.1 to 1.3 ha. Each site was surveyed on eight or nine nights from sunset to sunrise. Activity was recorded using bat detectors and thermal infrared imaging cameras simultaneously at the pond and at a vineyard site 80 m away. No comparisons were made to bat activity levels at natural pond sites.

(1) Vindigni M.A., Morris A.D., Miller D.A. & Kalcounis-Rueppell M.C. (2009) Use of modified water sources by bats in a managed pine landscape. *Forest Ecology and Management*, 258, 2056–2061.
(2) Stahlschmidt P., Pätzold A., Ressl L., Schulz R. & Brühl C.A. (2012) Constructed wetlands support bats in agricultural landscapes. *Basic and Applied Ecology*, 13, 196–203.

2.5 Retain or replace existing bat commuting routes

- We found no evidence for the effects of retaining or replacing the original commuting routes of bats lost due to agricultural land use change.

> **Background**
> Linear features provide connectivity for bats within open agricultural landscapes. Frey-Ehrenbold et al. (2013) found bat activity to be 1.4–2.8 times higher along linear features than in open farmland areas. One study in the UK found bats to be highly sensitive to the loss of field boundaries (Pocock & Jennings 2008). Another UK study highlights the importance of hedgerow trees for the use of linear features by bats (Boughey et al. 2011). We found no evidence for the effects of either retaining bat commuting routes within agricultural areas, or replacing them by planting. For evidence relating to creating commuting routes to divert bats see 'Threat: Transportation – Roads'.

Boughey K.L., Lake I.R., Haysom K.A. & Dolman P.M. (2011) Improving the biodiversity benefits of hedgerows: how physical characteristics and the proximity of foraging habitat affect the use of linear features by bats. *Biological Conservation*, 144, 1790–1798.

Frey-Ehrenbold A., Bontadina F., Arlettaz R. & Obrist M.K. (2013) Landscape connectivity, habitat structure and activity of bat guilds in farmland-dominated matrices. *Journal of Applied Ecology*, 50, 252–261.

Pocock M.J.O. & Jennings N. (2008) Testing biotic indicator taxa: the sensitivity of insectivorous mammals and their prey to the intensification of lowland agriculture. *Journal of Applied Ecology*, 45, 151–160.

Intensive farming

2.6 Convert to organic farming

- Evidence on whether organic farming benefits bats is mixed. Four replicated, paired, site comparison studies on farms in the UK[1, 2, 4] and Greece[3] had inconsistent results.

- Two studies in the UK[1, 2] found higher bat abundance and activity on organic farms than conventional farms. One of the studies[2] found more bat species on organic farms, and the other study[1] did not find a difference between farm types.

- One study in the UK[4] found no difference in bat abundance between fields in organic and conventional farms. A study on Zakynthos, Greece[3] found that a single yearly application of insecticide chemicals did not affect bat activity over traditional olive groves.

> **Background**
> Organic farming is an agricultural system that excludes the use of synthetic fertilizers and pesticides and relies on techniques such as crop rotation, compost and biological pest control. Organic standards are strictly regulated in many countries prohibiting the use of chemicals and providing recommendations for management to conserve biodiversity.

A replicated, paired, site comparison study in summer 2000 and 2002 on 24 pairs of farms in southern England and Wales, UK (1) found bat activity to be significantly higher over water habitats on organic farms than on conventional farms (447 vs. 144 total bat passes respectively). Bat activity did not differ significantly between farm types over pasture, arable or woodland habitats. Sixteen bat species were detected on organic farms and 11 on conventional farms but the difference was not significant. The activity of *Pipistrellus* and *Nyctalus* spp. did not differ significantly between farm types. *Myotis* spp. were recorded more on organic farms, and *Rhinolophus* spp. were only detected on organic farms with the majority in woodland habitats. Certified organic farms that had been established for one to two years were paired with nearby conventional farms of a comparable business, size, and number and area of habitat types. Two farms of each pair were sampled from June to September on consecutive nights. Selected habitats (pasture, arable land, water and woodland) found to be present on both farms of a pair were sampled. No details are given about the type or origin of water habitats sampled. Bat activity was recorded using bat detectors for ten minutes at three random points within each habitat at each site. Recordings were taken for an hour and a half from one hour after sunset. Analysis of habitat surveys showed that pairs of farms were comparable in all aspects except for hedgerow height, which was significantly higher on organic farms.

In a replicated, paired, site comparison study in summer 2002 and 2003 on 65 pairs of farms in England, UK (2) significantly more bat passes were recorded on organic farms than non-organic farms (abundance index 6–75% higher). Significantly more species were also recorded on organic farms (species density 8–65% higher). Organic farms of at least 30 ha of arable land were paired with nearby non-organic farms matched by crop type and cropping season. Bat surveys using bat detectors were conducted along 3 km triangular transects starting in a randomly chosen field on each farm between June and August in both years. Habitat data collected at all sites showed that organic farms had a higher density of hedgerows, a greater proportion of grassland than cropped cover, smaller fields and wider and taller hedgerows with fewer gaps than non-organic farms. The study looked at a variety of taxa and details about the levels of bat activity and bat species recorded are not given.

A replicated, paired, site comparison study in summer 2005 in olive *Olea europea* groves on Zakynthos island, Greece (3) found that bat activity did not differ significantly between six organic and six non-organic olive groves (average 0.8 vs. 1.1 bat passes/min respectively). Bat foraging activity and the activity of insect prey also did not differ significantly. Olive groves were similar in size, age, density of trees and altitude. Organic olive groves used organic pest control and no chemicals. Non-organic groves were treated with a yearly insecticide spray treatment. Bat and insect activity in both olive groves also did not differ significantly from that in six native woodland patches (oak and pine). Each site was sampled for three nights, with rotations between sites and habitat types. Bats were captured in mist nets and activity was recorded for an hour and a half from dusk in ten minute intervals rotating between four points within each site. Eleven bat species were detected.

A replicated, paired, site comparison study in summer 2003 on eight paired farms near Bristol, UK (4) did not find a significant difference in the abundance of common pipistrelles *Pipistrellus pipistrellus* between organic cereal fields and nearby conventionally farmed fields (total 96 vs. 152 bat passes respectively). Pairs of fields were matched to control for habitat variables and were sampled simultaneously. Each site was sampled on one night between May and August. Bat activity was recorded from 45 minutes after sunset using bat detectors for 20 minutes at four different points along a transect at each site. Two sample points were located 50 m into the field, and two within 1 m from the field boundary.

(1) Wickramasinghe L.P., Harris S., Jones G. & Vaughan, N. (2003) Bat activity and species richness on organic and conventional farms: impact of agricultural intensification. *Journal of Applied Ecology*, 40, 984–993.

(2) Fuller R.J., Norton L.R., Feber R.E., Johnson P.J., Chamberlain D.E., Joys A.C., Mathews F., Stuart R.C., Townsend M.C., Manley W.J., Wolfe M.S., Macdonald D.W. & Firbank L.G. (2005) Benefits of organic farming to biodiversity vary among taxa. *Biology Letters*, 1, 431–434.

(3) Davy C.M., Russo D. & Fenton M.B. (2007) Use of native woodlands and traditional olive groves by foraging bats on a Mediterranean island: consequences for conservation. *Journal of Zoology*, 273, 397–405.

(4) Pocock M.J.O. & Jennings N. (2008) Testing biotic indicator taxa: the sensitivity of insectivorous mammals and their prey to the intensification of lowland agriculture. *Journal of Applied Ecology*, 45, 151–160.

2.7 Introduce agri-environment schemes

- One replicated, paired, site comparison study in the UK[1] found lower bat activity and abundance of insect prey on farms participating in agri-environment schemes than on non-participating conventional farms.

Background

Agri-environment schemes provide farmers with financial incentives to manage their land in an environmentally friendly way. They promote the conservation of farmland, biodiversity and agro-ecosystems, and have been used in Europe, the USA, Canada and Australia.

Agri-environment schemes use many different specific interventions which may be beneficial to bats such as the protection and maintenance of archaeological features, traditional farm buildings and stone walls, the restoration and enhancement of high quality habitats such as woodland and hedgerows, and improvements to air and water quality. We did not find any studies looking at the individual effects of these interventions. One study examined the effects of agri-environment schemes overall.

A replicated, paired, site comparison study in summer 2008 on 18 pairs of farms in Scotland, UK (1) found that bat activity was significantly lower on farms participating in agrienvironment schemes (AES) than on non-participating conventional farms (790 vs. 1,175 total bat passes respectively). Bat foraging activity was also lower on AES farms (37 vs. 85 total feeding buzzes), as was the abundance of insect prey (5,039 vs. 10,193 total insects). Five bat species were recorded, with *Pipistrellus* spp. accounting for 98% of the total bat activity. Eighteen farms were selected which had been participating in the Scottish Rural Stewardship Scheme since 2004, and paired with nearby conventionally managed farms of a similar size and with similar farming activities. Each AES farm incorporated at least three management prescriptions likely to benefit bats (either the management of field margins, hedgerows, water margins or grasslands), and equivalent habitat features were selected on conventional farms. Each pair of farms was sampled once on the same night between June and September. Bat activity was recorded continuously from 45 minutes after sunset using bat detectors along transects 2.5–3.7 km in length. Each transect route incorporated the AES management prescriptions (or alternatives on conventional farms). Transects were of similar length and covered similar proportions of each habitat on paired farms. Nocturnal insects were sampled using three to four light traps per farm for fours after dusk.

(1) Fuentes-Montemayor E., Goulson D. & Park K.J. (2011) Pipistrelle bats and their prey do not benefit from four widely applied agri-environment management prescriptions. *Biological Conservation*, 144, 2233–2246.

13 Threat: Agriculture

2.8 Encourage agroforestry

- Six replicated, site comparison studies provide evidence as to whether agroforestry plantations benefit bats, with mixed results.
- Four studies, three in Mexico[1, 5, 6] and one in Costa Rica[4], found no difference in bat diversity, the number of bat species and/or bat abundance between coffee or banana agroforestry plantations and native rainforest fragments.
- One study in Mexico[3] found higher bat diversity and more bat species in native rainforest fragments than in coffee agroforestry plantations.
- One study in Costa Rica[4] found lower bat diversity and fewer bat species in native rainforest than cacao agroforestry plantations.
- One study in Brazil[2] found that landscape composition affected bat diversity in coffee agroforestry plantations. In areas where native rainforest dominated the landscape, bat diversity was higher in coffee agroforestry plantations than in native forest. Conversely, in areas where coffee agroforestry plantations dominated the landscape, bat diversity was lower in the coffee plantations than in native forest fragments.
- Three studies compared agroforestry plantations with different management intensities. Two studies in Mexico[3, 6] found lower numbers of insectivorous species and reduced activity of forest bat species, but similar or higher numbers of large fruit-eating bat species and increased activity of open space bat species in more intensively managed coffee agroforestry plantations. One study in Mexico[5] found a similar number of leaf-nosed bat species between three different management intensities on coffee agroforestry plantations, but lower bat capture rates in more intensively managed conventional plantations.

Background
Agroforestry is an integrated approach that involves growing crops or raising livestock under shade trees that are native tree species, remnants from cleared vegetation, or other crop trees.

Agroforestry farming provides a more complex habitat than conventional monoculture farming, and can support higher levels of biodiversity.

A replicated, site comparison study in 1998–1999 and 2001 in tropical forest and plantations in central Veracruz, Mexico (1) found a similar number of bat species in coffee agroforestry plantations and native rainforest fragments (ten species in each habitat with one species differing between them). Bats were sampled in three original rainforest fragments and three coffee agroforestry plantations. The agroforestry plantations consisted of coffee grown under a polyculture of shade trees with native tree species and planted fruit tree species. Each site was sampled using mist nets on two nights from June 1998 to May 1999, with an additional four nights of sampling from May to June 2001. Only bats of the families Phyllostomidae and Mormoopidae were counted.

In a replicated, site comparison study in 1998–2002 in tropical lowland rainforest and plantations in southern Bahia, Brazil (2) the diversity and number of bat species in cacao agroforestry plantations was found to differ between two different landscapes (Una and Ilhéus). In Una, where native rainforest is dominant with small patches of cacao agroforestry plantations, the number of bat species was higher in cacao plantations (Shannon-Wiener diversity index of 2.34, 39 species) than native forest (diversity index of 1.63, 21 species). In Ilhéus, where cacao agroforestry plantations are dominant with small patches of remnant rainforest, the number of bat species was lower in cacao plantations (diversity index of 1.61, 23 species) than native forest (diversity index of 1.82, 17 species). Shade cacao plantations in both landscapes were actively managed with a high degree of shade provided by mostly native canopy trees. Bat sampling was carried out using mist nets in three replicates of each habitat type from June 1998 to July 2001 in Una, and June to July 2002 in Ilhéus.

A replicated, site comparison study in 2004–2005 in six sites of montane rainforest and plantations in south-eastern Chiapas, Mexico (3) found higher bat diversity and more bat species in native rainforest than in coffee agroforestry plantations (Shannon-Wiener diversity index of 2.68 vs. 2.23–2.36, and 37 vs. 23–26 species, respectively). Coffee agroforestry plantations under different management regimes had a similar number of species, but differed in species composition. The number of insectivorous species was lower in plantations with high chemical use than those with low chemical use (two vs. six species respectively). The number of fruit-eating (13–15) and nectarivorous (2–5) species were similar between sites. The number of bat species was positively correlated with the number of vegetation layers, and the height and cover of trees. One site of native rainforest was sampled, and five sites on coffee agroforestry plantations with different levels of shade (original rainforest trees or one genus of leguminous shade trees) and chemical use (either none at all, organic compost, Thiodan, herbicide or fertilizer use). Plantations with the highest chemical input used all three chemical types. Bats were captured using mist nets during two nights at different locations at each site every two months from March 2004 to June 2005. Vegetation composition and structure were sampled at each site using a random sampling technique.

A replicated, site comparison study in 2002–2003 in tropical lowland forest and plantations in Talamanca, Costa Rica (4) found significantly higher bat diversity and a greater number of bat species in cacao agroforestry systems than in native forest (diversity index of 2.22 vs. 2.03, 15 vs. 13 species, respectively). Bat diversity and the number of bat species did not differ significantly between banana agroforestry plantations (diversity index of 2.19, 14 species) and native forest. Plantain monocultures were found to have significantly lower bat diversity and number of bat species (diversity index of 1.72, 10 species) than both agroforestry systems and native forest. Bat abundance did not vary significantly between forest, plantain monoculture, banana or cacao agroforestry systems (47, 83, 76 and 89 bats captured respectively). Both agroforestry systems were grown organically with a shade canopy of native trees or planted fruit and timber trees. Plantain monocultures were grown in patches of a similar size without shade and with the use of chemicals such as insecticides. Thirty-five sites were selected including 7 replicates each of forest, plantain monoculture and banana agroforestry, and 14 replicates of cacao agroforestry. Bats were sampled with mist nets for a total of 20 mist net hours per site per night from May 2002 to November 2003.

A replicated, site comparison study in 2006–2007 in tropical rainforest and coffee plantations in south-western Chiapas, Mexico (5) found a similar number of leaf-nosed bat species (family Phyllostomidae) and capture rate of bats in native rainforest fragments and traditional coffee agroforestry plantations (24 vs. 23 species respectively, both average 12 bats caught/mist net/hour). More intensively managed coffee plantations were found to have a similar number of bat species (22 species) but significantly lower bat capture rates (average 9 bats caught/mist net/hour). Bat species from all feeding guilds were found to decrease as management intensity increased, with the exception of large fruit-eating species which increased in proportion (from 30% in low management plantations to 48% in high management plantations). Bats were sampled in native forest fragments and coffee plantations with either low (traditional polyculture), moderate (commercial polyculture), or high (shade monoculture) management intensity. Sampling was conducted using mist nets over a total of 44 nights between November 2006 and August 2007.

A replicated, site comparison study in 2006–2007 in tropical rainforest and coffee plantations in south Chiapas, Mexico (6) found that the number of insectivorous bat species was not significantly different between native rainforest fragments and coffee agroforestry plantations with three different intensities of management (18 vs. 17–20 species respectively). The number of bat passes of forest bat species was similar between rainforest fragments (average 18 bat passes/night) and coffee plantations with low intensity management (average 21 bat passes/night), but was significantly lower in plantations with high intensity management (average 6 bat passes/night). Open space bat species showed the opposite pattern, with significantly more bat passes in coffee plantations with high intensity management (average three bat passes/night) than any of the other land use types (all average one bat pass/night). Bats were sampled in native forest fragments and coffee plantations with either low (traditional polyculture), moderate (commercial polyculture), or high (shade monoculture) management intensity. Sampling was conducted with mist nets and acoustic monitoring over a total of 44 nights between November 2006 and August 2007.

(1) Pineda E., Moreno C.E., Escobar F. & Halffter G. (2005) Frog, bat and dung beetle diversity in the cloud forest and coffee agroecosystems of Veracruz, Mexico. *Conservation Biology*, 19, 400–410.
(2) Faria D., Laps R.R., Baumgarten J. & Cetra M. (2006) Bat and bird assemblages from forests and shade cacao plantations in two contrasting landscapes in the Atlantic Forest of southern Bahia, Brazil. *Biodiversity & Conservation*, 15, 587–612.
(3) Estrada C.G., Damon A., Hernández C.S., Pinto S.L. & Núñez G.I. (2006) Bat diversity in montane rainforest and shaded coffee under different management regimes in southeastern Chiapas, Mexico. *Biological Conservation*, 132, 351–361.
(4) Harvey C.A. & Villalobos J.A.G. (2007) Agroforestry systems conserve species-rich but modified assemblages of tropical birds and bats. *Biodiversity and Conservation*, 16, 2257–2292.
(5) Williams-Guillén K. & Perfecto I. (2010) Effects of agricultural intensification on the assemblage of leaf-nosed bats (Phyllostomidae) in a coffee landscape in Chiapas, Mexico. *Biotropica*, 42, 605–613.
(6) Williams-Guillén K. & Perfecto I. (2011) Ensemble composition and activity levels of insectivorous bats in response to management intensification in coffee agroforestry systems. *PLoS ONE*, 6, e16502.

3 Threat: Energy production – wind turbines

Background
Renewable energy sources, such as wind power, have increased dramatically over the last few decades. Most wind energy development has been on commercial wind farms that have multiple large wind turbines with rotor diameters up to and over 100 m, each generating up to 2.3 MW. Smaller 'micro' wind turbines (which typically generate up to 50–100 kW) have also become increasingly popular, usually installed singly by homeowners on private land.

The evidence provided relates to large commercial wind turbines. We found no evidence for interventions relating to small 'micro' wind turbines.

Key messages

Modify turbine design to reduce bat fatalities
We found no evidence for the effects of modifying wind turbine design on bat fatalities.

Modify turbine placement to reduce bat fatalities
We found no evidence for the effects of altering the geographic location of wind turbines on bat fatalities.

Leave a minimum distance between turbines and habitat features used by bats
We found no evidence for the effects of creating buffers of a minimum distance between wind turbines and habitat features used by bats.

Deter bats from turbines using radar
A replicated, site comparison study in the UK found reduced bat activity in natural habitats in proximity to electromagnetic fields produced by radars. We found no evidence for the effects of installing radars on wind turbines on bats.

Deter bats from turbines using ultrasound
Five field studies at wind farms or pond sites (including one replicated, randomized, before-and-after trial), and one laboratory study, have all found lower bat activity or fewer bat deaths with ultrasonic deterrents than without.

Remove turbine lighting to avoid attracting bats
We found no evidence for the effects of removing lighting from wind turbines on bats.

Switch off turbines at low wind speeds to reduce bat fatalities
Three replicated, controlled studies in Canada and the USA have shown that reducing the operation of wind turbines at low wind speeds causes a reduction in bat fatalities.

17 Threat: Energy production – wind turbines

Automatically switch off wind turbines when bat activity is high
We found no evidence for the effects of automatically switching off wind turbines when high levels of bat activity are detected.

Close off nacelles on wind turbines to prevent roosting bats
We found no evidence for the effects of closing off access to nacelles on wind turbines to discourage roosting bats.

3.1 Modify turbine design to reduce bat fatalities

- We found no evidence for the effects of modifying wind turbine design on bat fatalities.

Background
Studies of patterns of bat fatalities at existing wind farms in Europe and the USA have shown that higher numbers of bats are killed at taller wind turbines (e.g. Barclay *et al.* 2007, Fielder *et al.* 2007, Rydell *et al.* 2010, Georgiakakis *et al.* 2012). Mortality was found to increase with rotor diameter in some studies (Rydell *et al.* 2010), but not in others (Barclay *et al.* 2007, Georgiakakis *et al.* 2012). We found no evidence for the effects of modifying wind turbine design on bat fatalities.

Barclay R.M.R., Baerwald E.F. & Gruver J.C. (2007) Variation in bat and bird fatalities at wind energy facilities: assessing the effects of rotor size and tower height. *Canadian Journal of Zoology*, 85, 381–387.
Fiedler J.K., Henry T.H., Tankersley R.D. & Nicholson C.P. (2007) *Results of Bat and Bird Mortality Monitoring at the Expanded Buffalo Mountain Windfarm, 2005*. Tennessee Valley Authority, Knoxville, USA.
Rydell J., Bach L., Dubourg-Savage M.-J., Green M., Rodrigues L. & Hedenström A. (2010) Bat mortality at wind turbines in northwestern Europe. *Acta Chiropterologica*, 12, 261–274.
Georgiakakis P., Kret E., Carcamo B., Doutau B., Kafkaletou-Diez A., Vasilakis D. & Papadatou E. (2012) Bat fatalities at wind farms in north-eastern Greece. *Acta Chiropterologica*, 14, 459–468.

3.2 Modify turbine placement to reduce bat fatalities

- We found no evidence for the effects of altering the geographic location of wind turbines on bat fatalities.

Background
Positioning wind turbines away from bat roosts, foraging areas and commuting or migration routes may reduce bat mortality. At wind farms in the USA, bat fatalities are often dominated by migratory species and are higher during autumn migration periods (e.g. Arnett *et al.* 2008, Baerwald & Barclay 2009, Piorkowski & O'Connell 2010). A review of reports in northwest Europe found higher fatality rates at wind farms located on forested hills than in flat, open farmland (Rydell *et al.* 2010). Spatial patterns of bat fatalities within wind farms in Europe and the USA have been found in some studies (Arnett *et al.* 2008, Baerwald & Barclay 2011, Georgiakakis *et al.* 2012) but not others (Arnett *et al.* 2008, Piorkowski & O'Connell 2010). We found no evidence for the effects of altering the geographic location of wind turbines on bat fatalities.

Arnett E.B., Brown W.K., Erickson W.P., Fiedler J.K., Hamilton B.L., Henry T.H., Jain A., Johnson G.D., Kerns J., Koford R.R., Nicholson C.P., O'Connell T.J., Piorkowski M.D. & Tankersley R.D. (2008)

Patterns of bat fatalities at wind energy facilities in North America. *The Journal of Wildlife Management*, 72, 61–78.

Baerwald E.F. & Barclay R.M.R. (2009) Geographic variation in activity and fatality of migratory bats at wind energy facilities. *Journal of Mammalogy*, 90, 1341–1349.

Piorkowski M.D. & O'Connell T.J. (2010) Spatial pattern of summer bat mortality from collisions with wind turbines in mixed-grass prairie. *The American Midland Naturalist*, 164, 260–269.

Rydell J., Bach L., Dubourg-Savage M.-J., Green M., Rodrigues L. & Hedenström A. (2010) Bat mortality at wind turbines in northwestern Europe. *Acta Chiropterologica*, 12, 261–274.

Baerwald E.F. & Barclay R M.R. (2011) Patterns of activity and fatality of migratory bats at a wind energy facility in Alberta, Canada. *The Journal of Wildlife Management*, 75, 1103–1114.

Georgiakakis P., Kret E., Carcamo B., Doutau B., Kafkaletou-Diez A., Vasilakis D. & Papadatou E. (2012) Bat fatalities at wind farms in north-eastern Greece. *Acta Chiropterologica*, 14, 459–468.

3.3   Leave a minimum distance between turbines and habitat features used by bats

- We found no evidence for the effects of creating buffers of a minimum distance between wind turbines and habitat features used by bats.

Background

This intervention involves leaving a minimum distance between wind turbines and bat roosts or habitat features to create a buffer zone. The EUROBAT guidance on bats and wind turbines recommends a minimum distance of 200 m between forest and wind turbines (Rodrigues *et al.* 2008), Natural England, UK recommends a minimum distance of 50 m from the turbine blade tip to the nearest bat habitat feature (Mitchell-Jones & Carlin 2012), and for micro turbines a minimum distance of 20 m has been recommended (Minderman *et al.* 2012).

Minderman J., Pendlebury C.J., Pearce-Higgins J.W. & Park K.J. (2012) Experimental evidence for the effect of small wind turbine proximity and operation on bird and bat activity. *PLoS ONE*, 7, e41177.

Mitchell-Jones, T & Carlin C. (2012) *Bats and Onshore Wind Turbines Interim Guidance*. Natural England Technical Information Note TIN051.

Rodrigues L., Bach L., Dubourg-Savage M.-J., Goodwin J. & Harbusch C. (2008) *Guidelines for Consideration of Bats in Wind Farm Projects*. EUROBATS Publication Series No. 3 (English version). UNEP/EUROBATS Secretariat, Bonn, Germany.

3.4   Deter bats from turbines using radar

- A replicated, controlled study in Scotland, UK[1] found that bat activity was reduced in proximity to electromagnetic fields produced by radars with fixed antennas. We found no evidence for the effects on bats of installing radars on wind turbines.

Background

It has been suggested that bats may avoid the radio frequency radiation associated with radar installations. There is evidence that bat activity is reduced in foraging habitats at air traffic control and weather radar sites with a high electromagnetic field (Nicholls & Racey 2007).

Nicholls B. & Racey P.A. (2007) Bats avoid radar installations; could electromagnetic fields deter bats from colliding with wind turbines? *PLoS ONE*, 2, e297.

A replicated, controlled study in summer 2009 across 20 wetland and woodland foraging sites in Scotland, UK (1) found that bats were significantly less active during experimental trials with fixed antenna radars emitting short pulse lengths and medium pulse lengths compared to control trials without radars (5,334 and 3,538 vs. 6,038 total bat counts respectively). Foraging rates (ratio of feeding buzzes to bat passes) were also lower with short (0.3) and medium (0.3) pulse signals from fixed antennas than in control trials (3.4). There was no significant difference in bat counts or foraging rates between control and experimental trials when a short pulse radar signal from a rotating antenna was used. On each night, a control and an experimental trial were carried out at one site. Bat activity was recorded for a period of 30 minutes (from 45 min after sunset) for each trial, and the order of trials was alternated between nights. Each of the three experimental treatments was tested once per site. Bat detectors were linked to count data loggers and placed at distances of 10, 20 and 30 m from the radar. Bat counts did not vary significantly between the three distances, and average counts were used. The species recorded were common pipistrelle *Pipistrellus pipistrellus*, soprano pipistrelle *Pipistrellus pygmaeus* and Daubenton's bat *Myotis daubentonii*. Radars were placed on a platform 2 m above ground level. Short pulse lengths were 0.08 μs/2100 Hz, and medium pulse lengths 0.3 μs/2,100 Hz.

(1) Nicholls B. & Racey P.A. (2009) The aversive effect of electromagnetic radiation on foraging bats – a possible means of discouraging bats from approaching wind turbines. *PLoS ONE*, 4, e6246.

3.5 Deter bats from turbines using ultrasound

- Five field studies (including one replicated, randomized, before-and-after trial), and one replicated, controlled laboratory study, have all found lower bat activity or fewer bat deaths with ultrasonic deterrents than without. The field studies were at wind farms[4, 5] or pond sites[2, 3, 6].

Background
Bats rely on ultrasound to echolocate for foraging and navigation. It has been suggested that broadcasting ultrasonic sounds at the frequency range which bats use for echolocation may act as a deterrent by interfering with their ability to perceive echoes.

A replicated, controlled laboratory experiment in 2006 at the University of Maryland, USA (1) found that six adult big brown bats *Eptesicus fuscus* flew significantly less through quadrants in a flight chamber when an acoustic deterrent emitted broadband white noise at frequencies from 12.5–112.5 kHz. With the deterrent active, fewer bat passes were recorded during feeding trials, with a mealworm tethered close to the deterrent (average 0.5 bat passes), and during non-feeding trials (average 0.4 bat passes), than in control trials with the deterrent switched off (average 1.6 bat passes feeding and 1.5 bat passes non-feeding). In non-feeding trials, bats also landed significantly less in quadrants with the deterrent active (2% of bats) than in control trials (22%). In feeding trials there was no significant difference in the frequency of bats landing in the quadrant between control and experimental trials. No bats successfully took a mealworm with the deterrent active but did so in 36% of trials when it was silent. Three bats were tested in non-feeding trials, and three bats previously trained to take a mealworm from a ceiling were tested in feeding trials. All bats were acclimatized to the testing chamber prior to the experiments. For each trial type, either 35 or 40 control trials were conducted (deterrent switched off) and 36 or 40 experimental trials (deterrent broadcasting noise). The acoustic deterrent was not tested in the field.

In a small, replicated, before-and-after study in the summer of 2006 at eight pond sites in California and Oregon, USA (2), bat activity was found to be lower when an ultrasonic deterrent was used. The average baseline activity without the deterrent (419 bat passes) was significantly higher than when the deterrent was active (238 bat passes). Bat passes were recorded visually using night vision video cameras at each site for a one hour period after sunset for two nights with no deterrent, and then on a third night with the deterrent activated.

In a replicated, before-and-after study in August and September 2007 at 6 pond sites in Arizona, California and Oregon, USA (3), the hourly bat activity rate was significantly lower with an acoustic deterrent emitting ultrasonic sound than when no deterrent was used (average 19 vs. 288 bat passes/night respectively). Bat activity was recorded visually using night vision video cameras for one hour per night, with at least two nights for baseline activity followed by five to seven days of continuous ultrasound treatment at each site. The deterrent broadcast continuous broadband ultrasound at 20–80 kHz, with a range of 12–15 m. The authors note that rapid attenuation of ultrasound in air may limit the effectiveness of acoustic deterrents as mitigation.

A small, paired, site comparison study in August 2007 on a wind farm in an agricultural area of New York, USA (4) found mixed effects on bat activity when an ultrasound bat deterrent was used. Significantly fewer bats were observed over ten consecutive nights at a turbine with an ultrasonic deterrent fitted (average 13 bat passes/night) than at a matched control turbine without a deterrent (average 24 bat passes/night). No significant difference was found in bat activity when this was repeated with a second matched pair. Bat activity was observed for 3.6 hours after sunset using thermal IR imaging cameras for both trials. The deterrent broadcast random pulses of broadband ultrasound from 20–80 kHz, with a range of up to 20 m.

A replicated, randomized, controlled study in 2009–2010, with a before-and-after trial in the second year, in Pennsylvania, USA (5) found varying effects on bat mortality of six species when an ultrasonic deterrent was used at a wind farm within a forested area. In 2009, 21–51% fewer bats were found to be killed per deterrent turbine than control turbine (with average fatality rates of 12 bats/turbine with deterrent vs. 18 bats/turbine control) In the 2010 before-and-after trial, between 2% more and 64% fewer bats were killed at deterrent turbines than at control turbines when accounting for inherent differences found between control and deterrent turbines in the 'before' trial. In both years, ten randomly selected wind turbines were fitted with acoustic deterrent devices, and fifteen randomly selected turbines without the device were used as controls. In 2009, daily carcass searches (corrected for searcher efficiency and carcass removal) were conducted from August to October. In 2010, the before-and-after trial was conducted with daily carcass searches from May to July before the deterrent treatment was used, followed by daily searches from July to October with the deterrent active. The deterrent emitted continuous ultrasonic broadband noise at 20–100 kHz, with the range likely to be up to 5–10 m for the upper frequency ranges (70–100 kHz) but highly dependent on air temperature, pressure and humidity.

A small, replicated, before-and-after trial of bat activity at four forest pond sites in July 2009 in West Virginia, USA (6) found that bat activity was reduced by 17% when an ultrasonic deterrent was used (average 345 bats passes/night reduced to 286). Each pond was monitored for four consecutive nights for baseline activity levels. The deterrent was deployed at two of the pond sites chosen at random for a further three nights, and then at the remaining two ponds for three nights and activity recorded. The deterrent broadcast ultrasound noise at 26–74 kHz. The range of the deterrent is not given, but was large enough to encompass each pond site. The authors highlight the need for ultrasound emission to reach beyond the wind turbine blade lengths.

(1) Spanjer G.R. (2006) *Responses of the Big Brown Bat, Eptesicus fuscus, to a Proposed Acoustic Deterrent Device in a Lab Setting*. A report submitted to the Bats and Wind Energy Cooperative and the Maryland Department of Natural Resources. Bat Conservation International, Austin, Texas, USA.
(2) Szewczak J.M. & Arnett E. (2006) *Preliminary Field Test Results of an Acoustic Deterrent with the Potential to Reduce Bat Mortality from Wind Turbines*. An investigative report submitted to the Bats and Wind Energy Cooperative. Bat Conservation International, Austin, Texas, USA.
(3) Szewczak J.M. & Arnett E.B. (2008) *Field Test Results of a Potential Acoustic Deterrent to Reduce Bat Mortality from Wind Turbines*. An investigative report submitted to the Bats and Wind Energy Cooperative. Bat Conservation International, Austin, Texas, USA.
(4) Horn J.W., Arnett E.B., Jensen M. & Kunz T.H. (2008) *Testing the Effectiveness of an Experimental Bat Deterrent at the Maple Ridge Wind Farm*. A report submitted to The Bats and Wind Energy Cooperative. Bat Conservation International, Austin, Texas, USA.
(5) Arnett E.B., Hein C.D., Schirmacher M.R., Baker M., Huso M.M.P. & Szewczak J.M. (2011) *Evaluating the Effectiveness of an Ultrasonic Acoustic Deterrent for Reducing Bat Fatalities at Wind Turbines*. A final report submitted to the Bats and Wind Energy Cooperative. Bat Conservation International, Austin, Texas, USA.
(6) Johnson J.B., Ford W.M., Rodrigue J.L. & Edwards J.W. (2012) *Effects of Acoustic Deterrents on Foraging Bats*. Research Note NRS-129. Newtown Square, PA: U.S. Department of Agriculture, Forest Service, Northern Research Station.

3.6 Remove turbine lighting to avoid attracting bats

- We found no evidence for the effects of removing lighting from wind turbines on bats.

Background
It has been suggested that lights placed on wind turbines may attract insects and foraging bats and increase the risk of collision. Several studies in the USA (e.g. Johnson *et al.* 2004, Jain *et al.* 2010, Baerwald & Barclay 2011) found that bat fatalities were not increased at turbines lit with aviation lighting in comparison to unlit turbines. We found no evidence for the effects of removing lighting from wind turbines on bats.

Johnson G.D., Perlik M.K., Erickson W.P. & Strickland M.D. (2004) Bat activity, composition, and collision mortality at a large wind plant in Minnesota. *Wildlife Society Bulletin*, 32, 1278–1288.
Jain A.A., Koford R.R., Hancock A.W. & Zenner G.G. (2010) Bat mortality and activity at a northern Iowa wind resource area. *The American Midland Naturalist*, 165, 185–200.
Baerwald E.F. & Barclay R.M.R. (2011) Patterns of activity and fatality of migratory bats at a wind energy facility in Alberta, Canada. *The Journal of Wildlife Management*, 75, 1103–1114.

3.7 Switch off turbines at low wind speeds to reduce bat fatalities

- Three replicated, controlled studies from Canada and the USA have all found fewer bat fatalities when turbine operation is reduced at low wind speeds. Turbines were either shut down at low wind speed[1], turbine cut-in speed was increased[2, 3], or the angle of the rotor blades was altered[2].

Background
Most wind turbines operate by a 'cut-in' wind speed at which the turbine begins to generate electricity and the blades can move at a maximum rotation speed. However,

3.7 Switch off turbines at low wind speeds to reduce bat fatalities

the blades can still rotate at lower wind speeds when electricity is not being generated. As bat fatality rates have been found to be greater at low wind speeds (see for example Kerns et al. 2005, Horn et al. 2008, Rydell et al. 2010), it has been suggested that wind turbines should be shut down when wind speeds are low, and turbine cut-in speeds should be increased.

Kerns J., Erickson W.P. & Arnett E.B. (2005) Bat and bird fatality at wind energy facilities in Pennsylvania and West Virginia. In Arnett E. B. (ed.) *Relationships Between Bats and Wind Turbines in Pennsylvania and West Virginia: An Assessment of Bat Fatality Search Protocols, Patterns of Fatality, and Behavioral Interactions with Wind Turbines*. A final report submitted to the Bats and Wind Energy Cooperative. Bat Conservation International, Austin, Texas, USA. pp. 24–95.
Horn J.W., Arnett E.B. & Kunz T.H. (2008) Behavioral responses of bats to operating wind turbines. *The Journal of Wildlife Management*, 72, 123–132.
Rydell J., Bach L., Dubourg-Savage M.-J., Green M., Rodrigues L. & Hedenström A. (2010) Bat mortality at wind turbines in northwestern Europe. *Acta Chiropterologica*, 12, 261–274.

In a small, replicated, controlled study in 2005 in an agricultural area of Alberta, Canada (1) the shutdown of turbines in low wind speeds significantly reduced bat mortality. The number of bat carcasses recovered by searchers was significantly lower (64, 40% of total) at turbines that were shut down below wind speeds of 4 m/s than at those left on (157, 49% of total). Carcass searches were conducted weekly throughout August and September with standardized transects covering a 140 m^2 area around each turbine. Carcass recovery rates for the five species found were adjusted for searcher efficiency. Throughout August all turbines operated as normal (with similar mortality rates across all turbines). In September odd numbered turbines (20 out of 39) were shut down in wind speeds below 4 m/s. The experiment did not adversely impact the electricity generation of the turbines.

A replicated, randomized, controlled study in 2006 and 2007 in an agricultural area of Alberta, Canada (2) found that fatalities of hoary bats *Lasiurus cinereus* and silver-haired bats *Lasionycteris noctivagans* were reduced when two different operational changes were made to wind turbines to reduce rotor rotation at low wind speeds. Bat fatality rates were significantly lower at experimental turbines with increased cut-in speed or altered blade angle (both average 8 bats/turbine) than control turbines (average 19 bats/turbine). In both years, from August to September, carcass searches were conducted weekly along spiral transects up to 52 m around each of the 39 turbines. Bat fatality rates were corrected for searcher efficiency and scavenger removal. In 2006, no operational changes were made and fatality rates were not found to differ between turbines. In 2007, 15 randomly chosen turbines were altered by increasing the cut-in wind speed to 5.5 m/s, and 6 randomly chosen turbines were altered by changing the pitch angle of the rotor blades. The remaining control turbines were left unaltered. Both operational changes caused turbines to remain motionless at low wind speeds and bat fatality rates did not differ significantly between them. Turbines with increased cut-in speed caused a greater reduction in electricity generation than those with altered blade angles.

In a replicated, randomized, controlled study in 2008 and 2009 in a forested area of Pennsylvania, USA (3) average nightly bat fatalities were reduced when the cut-in wind speed of turbines was increased. In 2008 and 2009, bat fatalities were significantly higher at fully operational turbines (both years average 2 bats/turbine/night) than curtailed turbines (0.3 and 0.7 bats/turbine/night at 5 m/s cut-in speed, 0.5 and 0.6 bats/turbine/night at 6.5 m/s cut-in speed). Each night from July to October in both years, 12 out of a total of 23 turbines were randomly selected and assigned to one of three treatments: fully operational with a cut-in speed of 3.5 m/s, or curtailed with either a cut-in speed of 5 m/s or 6 m/s. There was

23 Threat: Energy production – wind turbines

no difference in the number of fatalities between the two curtailed turbine treatments. Daily carcass searches were conducted along transects in 120 x 126 m rectangular plots centred on each turbine, and fatality rates were corrected for searcher efficiency and scavenger removal. Six species were killed at the site with hoary bats *Lasiurus cinereus*, silver-haired bats *Lasionycteris noctivagans* and eastern red bats *Lasiurus borealis* being the most common. If applied to the entire wind farm, the operational changes made from July to October were projected to create an annual power output loss of 0.3% when cut-in speeds are increased to 5 m/s, and 1% when cut-in speeds are increased to 6 m/s.

(1) Brown W.K. & Hamilton B.L. (2006) *Monitoring of Bird and Bat Collisions with Wind Turbines at the Summerview Wind Power Project, Alberta, 2005–2006*. Vision Quest Windelectric, Calgary, Alberta, Canada.
(2) Baerwald E.F., Edworthy J., Holder M. & Barclay R.M.R. (2009) A large-scale mitigation experiment to reduce bat fatalities at wind energy facilities. *The Journal of Wildlife Management*, 73, 1077–1081.
(3) Arnett E.B., Huso M.M.P., Schirmacher M.R. & Hayes J.P. (2010) Altering turbine speed reduces bat mortality at wind-energy facilities. *Frontiers in Ecology and the Environment*, 9, 209–214.

3.8 Automatically switch off wind turbines when bat activity is high

- We found no evidence for the effects of automatically switching off wind turbines when high levels of bat activity are detected.

Background
This intervention involves the use of automatic bat registration systems to monitor bat activity and shut down operation of wind turbines when bat activity reaches a predetermined 'high' level.

3.9 Close off nacelles in wind turbines to prevent roosting bats

- We found no evidence for the effects of closing off access to nacelles on wind turbines to discourage roosting bats.

Background
Nacelles are the housing units at the top of wind turbines which contain the generator, electronic controls and inner workings of the turbine. They have warm, hollow spaces, and bats have been observed roosting in nacelles on offshore wind turbines in Scandinavia (Ahlén *et al.* 2009). Closing access to nacelles in wind turbines to prevent bats from roosting may reduce the risk of bat collisions with turbine blades.

Ahlén I., Baagøe H.J. & Bach L. (2009) Behavior of Scandinavian bats during migration and foraging at sea. *Journal of Mammalogy*, 90, 1318–1323.

4 Threat: Energy production – mining

Background

Abandoned mines are often used as roosting sites for cave-dwelling bats as they provide stable microclimates and shelter. However, abandoned mines can be hazardous to members of the public and are often closed and reclaimed by filling in, sealing, blasting or gating.

Key messages

Legally protect bat hibernation sites in mines from reclamation
We found no evidence for the effects of legally protecting bat hibernation sites in abandoned mines from reclamation.

Provide artificial hibernacula to replace roosts lost in reclaimed mines
We found no evidence for the effects of providing artificial hibernacula to replace bat hibernation sites lost during the reclamation of abandoned mines.

Relocate bats from reclaimed mines to new hibernation sites
We found no evidence for the effects of moving bats to new hibernation sites when existing hibernacula in abandoned mines are lost due to reclamation.

For interventions relating to **cave gating** and human disturbance of bats at abandoned mines see 'Threat: Human disturbance – Use cave gates to restrict public access'.

4.1 Legally protect bat hibernation sites in mines from reclamation

- We found no evidence for the effects of legally protecting bat hibernation sites in abandoned mines from reclamation.

4.2 Provide artificial hibernacula to replace roosts lost in reclaimed mines

- We found no evidence for the effects of providing artificial hibernacula to replace bat hibernation sites lost during the reclamation of abandoned mines.

4.3 Relocate bats from reclaimed mines to new hibernation sites

- We found no evidence for the effects of moving bats to new hibernation sites when existing hibernacula in abandoned mines are lost due to reclamation.

5 Threat: Transportation and service corridors

Background

Roads have been shown to have a negative impact on bats, acting as a barrier to movement and causing direct mortality due to collisions with vehicles (Lesinski 2007, Kerth & Melber 2009, Berthinussen & Altringham 2012). The habitat surrounding roads may also become unsuitable for bats due to light, noise and chemical pollution.

We have found no evidence for the effects of other transportation, such as rail, on bats and the following information relates to roads only.

The ultimate function of road crossing structures for bats is to increase road permeability and reduce mortality so as to maintain bat populations close to roads. We found no evidence to show that crossing structures either increase permeability or maintain bat populations in proximity to roads. We found evidence that some crossing structures are used by bats. Few crossing structures were used by a sufficient proportion of crossing bats to suggest they would be effective at maintaining bat populations.

Lesinski G. (2007) Bat road casualties and factors determining their number. *Mammalia*, 71, 138–142.
Kerth G. & Melber M. (2009) Species-specific barrier effects of a motorway on the habitat use of two threatened forest-living bat species. *Biological Conservation*, 142, 270–279.
Berthinussen A. & Altringham J. (2012) The effect of a major road on bat activity and diversity. *Journal of Applied Ecology*, 49, 82–89.

Key messages – Roads

Install underpasses as road crossing structures for bats
Four studies (two replicated) in Germany, Ireland and the UK found varying proportions of bats to be using existing underpasses below roads and crossing over the road above.

Install overpasses as road crossing structures for bats
One replicated, site comparison study in Ireland did not find more bats using over-motorway routes than crossing over the road below.

Install bat gantries or bat bridges as road crossing structures for bats
One replicated, site comparison study in the UK found fewer bats using bat gantries than crossing the road below at traffic height.

Install green bridges as road crossing structures for bats
We found no evidence concerning the proportions of bats using green bridges, or their effects on bat populations.

Install hop-overs as road crossing structures for bats
We found no evidence to show that hop-overs help bats to cross roads safely.

Divert bats to safe crossing points with plantings or fencing

We found no evidence for the effects of diverting bats to safe road crossing points. One controlled, before-and-after study in Switzerland found that a small proportion of lesser horseshoe bats within a colony flew along an artificial hedgerow to commute.

Deter bats with lighting

We found no evidence for the effects of deterring bats from roads with lighting.

Replace or improve habitat for bats around roads

We found no evidence for the effects of habitat improvements around roads.

5.1 Install underpasses as road crossing structures for bats

- Four studies (two replicated) in Germany, Ireland and the UK found bats to be using existing underpasses below roads and crossing over the road above in varying proportions[1, 2, 3, 4]. None of the underpasses tested were purpose-built for bats.

Background

Underpasses may guide bats safely under roads. They have the potential to reduce the number of bats killed by traffic and increase the permeability of roads for bats to maintain connectivity across the landscape. There is evidence that an unknown proportion of bats of various species use underpasses (e.g. Bach *et al.* 2004, Boonman 2011). Several studies report the proportion of bats that are either using existing underpasses to cross roads safely, or are crossing the road above at risk of collision with traffic. The underpasses studied were built to carry minor roads, paths or water. We found no studies on underpasses that have been purpose-built for bats.

Bach L., Burkhardt P. & Limpens H. (2004) Tunnels as a possibility to connect bat habitats. *Mammalia*, 68, 411–420.

Boonman M. (2011) Factors determining the use of culverts underneath highways and railway tracks by bats in lowland areas. *Lutra*, 54, 3–16.

In a replicated study in May–September from 2004 to 2007 in a forested area of northern Bavaria, Germany (1) seven bat species were caught in two underpasses below a road. Two of the captured species (barbastelle bats *Barbastella barbastellus* and Natterer's bats *Myotis nattereri*) were caught more often in underpasses (29 and 28 bats/mist net/night respectively) than at other sites within the forest (2 and 4 bats/mist net/night respectively). Only 3 of 34 radio-tracked Bechstein's bats *Myotis bechsteinii* crossed the motorway, all using one underpass (36 crossings). Five of six radio-tracked barbastelle bats crossed the motorway either through an underpass (16 crossings) or over the motorway (21 crossings at 6 different sites). Mist netting was carried out for 153 nights at 12 sites within the forest. Adult female bats were radio-tracked for at least three full consecutive nights. The motorway has four to five lanes and carries an average of 84,000 vehicles per day. Three underpasses are located within the forested section (width 10 or 5 m, height 4.5 m, length 30 m).

A replicated, site comparison study in May–September 2008 at 25 under- and over-motorway crossing routes in agricultural and woodland habitat in southern Ireland (2) found that bats used under-motorway routes (rivers which had been bridged over by the road or underpasses) more than over-motorway routes (bridges over the road or severed mature treelines level or above road height). More bats (excluding Leisler's bats *Nyctalus leisleri* which flew over the majority of structures) were recorded in under-motorway routes (4,692 bat passes at river bridges and 662 bat passes at underpasses) than over the road above (96 bat passes above river bridges and 45 bat passes above underpasses). Activity was greater (by > 10%) at under-motorway crossing routes than in adjacent habitats. Bat activity was recorded with bat detectors at six replicates of each type of crossing route (and an additional underpass) on two nights from ten minutes before to two hours after dusk. Recordings were made above and below the structures and simultaneously at adjacent linear features located 5–15 m from the motorway. The common pipistrelle *Pipistrellus pipistrellus*, soprano pipistrelle *Pipistrellus pygmaeus* and *Myotis* spp. made up 90% of the recordings. Brown long-eared bats *Plecotus auritus* and Leisler's bats *Nyctalus leisleri* were also detected. The motorway was opened in sections from 1992 to 2006, and has four lanes carrying an average of 20,000 vehicles per day.

A study in summer 2009 and 2010 at three underpasses below a motorway in agricultural and woodland habitat in Ennis, west Ireland (3) found different bat species using different sized underpasses. Lesser horseshoe bats *Rhinolophus hipposideros*, Natterer's bats *Myotis nattereri* and brown long-eared bats *Plecotus auritus* were the only species to fly through two narrow drainage pipes. All species flew through a larger underpass (except Leisler's bats *Nyctalus leisleri*, which always flew over the road). *Pipistrellus* spp. were detected most frequently in the large underpass (average 77 bat passes/night) but 18% flew over the road above (average 17 bat passes/night). The majority of brown long-eared bats, lesser horseshoe bats and *Myotis* spp. were recorded in the large underpass (total 60, 58 and 30 bat passes over 16 nights, respectively) with a small proportion recorded over the road above (total 3, 1 and 1 bat passes in 16 nights, respectively). Acoustic recordings were made in two long (43 m and 91 m) and narrow (approximately 1 x 1.4 m) drainage pipes and a large underpass (6 x 17 m, 26 m long) beneath a motorway. The underpass was surveyed for 16 full nights in May 2009, with simultaneous recordings above and below the underpass. The drainage pipes were surveyed for 17 full nights in August and September 2010. The motorway was constructed in 2007, with 4 traffic lanes carrying an average of 11,000 vehicles per day.

A study in June and July 2010 at three underpasses at roads within rural agricultural habitat in Cumbria, UK (4) found significantly more bats flying through one of the underpasses in preference to flying over the road above in the path of traffic (864 vs. 32 total bats respectively). The underpass (30 m length x 6 m width x 5 m height) was located on a pre-construction commuting route and bats were not required to alter their flight height or direction to pass through it. At the other two underpasses (15 m length x 5 m width x 2.5 m height, and 30 m length x 6 m width x 3 m height) fewer bats flew through them (39 and 11 total bats respectively) and more bats crossed the road above at traffic height (751 and 22 total bats respectively). Both underpasses were not located on pre-construction commuting routes, but attempts had been made to divert bats towards them with planting. Observations and recordings of bat calls were made during ten 90 minute surveys (five at sunset and five at sunrise) at each crossing point. The common pipistrelle *Pipistrellus pipistrellus*, soprano pipistrelle *Pipistrellus pygmaeus*, brown long-eared bat *Plecotus auritus* and *Myotis* spp. were included in the study. All roads had two or three lanes of traffic carrying an average of 12,000–17,000 vehicles per day.

(1) Kerth G. & Melber M. (2009) Species-specific barrier effects of a motorway on the habitat use of two threatened forest-living bat species. *Biological Conservation* 142, 270–279.
(2) Abbott I.M., Butler F. & Harrison S. (2012) When flyways meet highways – the relative permeability of different motorway crossing sites to functionally diverse bat species. *Landscape and Urban Planning*, 106, 293–302.
(3) Abbott I.M., Harrison S. & Butler F. (2012) Clutter-adaptation of bat species predicts their use of under-motorway passageways of contrasting sizes – a natural experiment. *Journal of Zoology*, 287, 124–132.

(4) Berthinussen A. & Altringham J. (2012b) Do bat gantries and underpasses help bats cross roads safely? *PLoS ONE*, 7, e38775.

5.2 Install overpasses as road crossing structures for bats

- One replicated, site comparison study in Ireland[1] found that more bats crossed over the road below than used over-motorway routes (such as overpasses).

> **Background**
> Overpasses, such as bridges built for pedestrians or vehicles, may help to guide bats safely over roads. This would both reduce the number of bats killed on roads, and increase the permeability of roads for bats to maintain connectivity across the landscape. We found no evidence to show that overpasses assist a significant proportion of bats to cross roads safely. One study reports the proportion of bats either crossing safely above overpasses or crossing over the road below.

A replicated, site comparison study in May–September 2008 at 25 under- and over-motorway crossing routes in agricultural and woodland habitat in southern Ireland (1) found that bats used over-motorway routes (bridges over the road or severed mature treelines level or above road height) less than under-motorway routes (rivers which had been bridged by the road or underpasses). At over-motorway routes, 50% of bat passes were recorded over the road below the crossing structure in the path of traffic (excluding Leisler's bats *Nyctalus leisleri*, which always flew over structures). Activity was lower (by > 10%) at over-motorway crossing routes than in adjacent habitats. Bat activity was recorded with bat detectors at six replicates of each type of crossing route (and an additional underpass) on two nights from ten minutes before to two hours after dusk. Recordings were made above and below the structures and also simultaneously at adjacent linear features located 5–15 m from the motorway. The common pipistrelle *Pipistrellus pipistrellus*, soprano pipistrelle *Pipistrellus pygmaeus* and *Myotis* spp. made up 90% of the recordings. Brown long-eared bats *Plecotus auritus* and Leisler's bats were also detected. The motorway was opened in sections from 1992 to 2006, and has four lanes carrying an average of 20,000 vehicles per day.

(1) Abbott I.M., Butler F. & Harrison S. (2012) When flyways meet highways – the relative permeability of different motorway crossing sites to functionally diverse bat species. *Landscape and Urban Planning*, 106, 293–302.

5.3 Install bat gantries or bat bridges as road crossing structures for bats

- One replicated, site comparison study in the UK[1] found fewer bats using bat gantries to safely cross roads than crossing the road at traffic height.

> **Background**
> Bat gantries, or bridges, are purpose-built structures designed to act as linear features that will guide echolocating bats over roads at a safe height above traffic. They typically consist of wood or metal pylons erected on either side of the road with wires, netting or other material spanning the road between them. The aim is to both reduce the number

of bats killed on roads, and increase the permeability of roads to maintain connectivity for bats across the landscape. We found no evidence for the effects of bat gantries or bridges to help a significant proportion of bats to cross roads safely and therefore to help maintain bat populations in proximity to roads. One study looks at the use of wire bat gantries by bats, and the proportion of bats crossing safely using the structure or crossing unsafely over the road below.

A replicated, site comparison study in June and July 2010 at four bat gantries (or bridges) on roads within rural agricultural habitat in northern England, UK (1) found significantly fewer bats using the bat gantries to safely cross roads than crossing below at traffic height. The number of bats using the gantries (flying within 2 m of them) to safely cross the roads at each site was 2, 5, 11 and 24 bats. The number of bats crossing at traffic height below the gantries was 81, 10, 43 and 751 bats at each site respectively. Comparable numbers of bats also crossed the roads at nearby commuting routes that did not have crossing structures in place, with the majority of bats crossing at traffic height. Observations and recordings of bat calls were made during ten 90-minute surveys (five at sunset and five at sunrise) at each crossing point (except for one which was only surveyed at dusk). The common pipistrelle *Pipistrellus pipistrellus*, soprano pipistrelle *Pipistrellus pygmaeus*, brown long-eared bat *Plecotus auritus* and *Myotis* spp. were included in the study. All roads had two or three lanes of traffic carrying an average of 12,000–17,000 vehicles per day. Bat gantries were of a similar design (height 6–9 m, width 2 m) with two or three pairs of wires spanning the road (20–30 m) and plastic spheres attached at 2 m intervals.

(1) Berthinussen A. & Altringham J. (2012b) Do bat gantries and underpasses help bats cross roads safely? *PLoS ONE*, 7, e38775.

5.4 Install green bridges as road crossing structures for bats

- We found no evidence concerning the proportions of bats using green bridges, or their effects on bat populations.

Background
Green bridges are bridges over roads that are covered in vegetation and usually planted with hedgerows and trees. They have been built in Europe and North America as mitigation measures usually to guide larger mammals, such as deer, safely across wide roads. A study in Germany found ten bat species using eight green bridges to fly over a road, and also to forage (Bach & Müller-Stiess 2005). Although we have evidence that bats will use green bridges, we do not know what proportion of crossing bats are using them or if they are effective in maintaining bat populations in proximity to roads.

Bach L. & Müller-Stiess H. (2005) Fachbeitrag Fledermäuse an ausgewählten Grünbrücken. Effizienzkontrolle von Wildtierpassagen in Baden-Württemberg (FE 02.220/2002/LR) In Georgii B., Peters-Ostenberg E., Henneberg M., Herman M., Müller-Stiess H. & Bach L. (2007) *Nutzung von Grünbrücken und anderen Querungsbauwerken durch Säugetiere*. Gesamtbericht zum Forschungs- und Entwicklungsvorhaben 02.247/2002LR.

5.5 Install hop-overs as road crossing structures for bats

- We found no evidence for the effects of hop-overs as road crossing structures for bats.

> **Background**
> A hop-over consists of tall vegetation planted on either side of a road with overhanging branches that create a continuous canopy over the road gap. The aim is to guide bats across roads at a safe height. While there is evidence that bats will cross roads at greater heights in the presence of high canopy cover or roadside embankments (Russell et al. 2009, Berthinussen & Altringham 2012b), we found no evidence for the effectiveness of hop-overs in guiding bats safely over roads and maintaining local bat populations.

Berthinussen A. & Altringham J. (2012b) Do bat gantries and underpasses help bats cross roads safely? *PLoS ONE*, 7, e38775.
Russell A.L., Butchkoski C.M., Saidak L. & McCracken G.F. (2009) Road-killed bats, highway design, and the commuting ecology of bats. *Endangered Species Research*, 8, 49–60.

5.6 Divert bats to safe crossing points with plantings or fencing

- We found no evidence for the effects of diverting bats to safe road crossing points. One controlled, before-and-after study in Switzerland[1] found that a small proportion of lesser horseshoe bats within a colony flew along an artificial hedgerow to commute.

> **Background**
> Linear features such as hedgerows and treelines provide important commuting routes for bats (e.g. Verboom & Huitema 1997, Limpens & Kapteyn 1991). Roads can fragment these commuting routes, cutting off important habitat. It has been suggested that bats may be diverted from original commuting routes to crossing structures or safe crossing places along roads by planting treelines or hedgerows, or installing fences. Berthinussen & Altringham (2012b) found that although diverted bats were not recorded directly, very few bats used two underpasses where attempts had been made to divert bats to them with plantings. Conversely, very high numbers of bats were found using an underpass constructed on an original flight path.

Berthinussen A. & Altringham J. (2012b) Do bat gantries and underpasses help bats cross roads safely? *PLoS ONE*, 7, e38775.
Limpens H.J. & Kapteyn K. (1991) Bats, their behaviour and linear landscape elements. *Myotis*, 29, 39–48.
Verboom B. & Huitema H. (1997) The importance of linear landscape elements for the pipistrelle *Pipistrellus pipistrellus* and the serotine bat *Eptesicus serotinus*. *Landscape Ecology*, 12, 117–125.

A controlled, before-and-after study in July to September 2003 at one site in Giswil, Switzerland (1) found that significantly more lesser horseshoe bats *Rhinolophus hipposideros* exiting from one side of a nearby roost flew in a particular direction after an artificial hedgerow had been installed for over two weeks (average 3% of bats/phase before, 10% of bats/phase after). A maximum of 20% of bats (approximately 20 out of 100) exiting the roost on the same side flew along the hedgerow in one night in the fourth phase of the artificial hedgerow being in place. Bats flying towards the artificial hedgerow were found to emerge earlier from the roost and return later than bats using other flight routes, and were out of the roost for up to four minutes longer. The experiment was split into phases of four to five nights, with one phase each for before and after control periods, and six experimental phases. Bat activity was monitored with bat detectors and infrared video recordings for at least 50 minutes at sunset and sunrise for 39 nights. The artificial hedgerow consisted of a 200 m long linear structure of native hedgerow plants in containers positioned through open farmland, connecting the bat roost and the main forested foraging habitat used by the bats. The bat roost was estimated to contain 260 individuals. The structure was 1 m wide and 1.5–2 m high. Camouflage netting was added to a 90 m section of the structure for the last two experimental phases, which increased the width to 2 m. With this addition, average numbers of bats using the structure over each phase (9 and 12%) remained similar to the previous phase with only the hedgerow in place (10%).

(1) Britschgi A., Theiler A. & Bontadina F. (2004) Wirkungskontrolle von Verbindungsstrukturen. Teilbericht innerhalb der Sonderuntersuchung zur Wochenstube der Kleinen Hufeisennase in Friedrichswalde-Ottendorf / Sachsen. Unveröffentlichter Bericht, ausgeführt von BMS GbR, Erfurt & SWILD, Zürich im Auftrage der DEGES, Berlin.

5.7 Deter bats with lighting

- We found no evidence for the effects of deterring bats from roads with lighting.

Background
Some bat species avoid lit areas (e.g. Stone *et al.* 2009, 2012), and it has been suggested that strategically placed lighting around roads may be used to deter bats from unsafe crossing points and divert them to safe crossing points.
  See also 'Threat: Pollution – Light and noise pollution' for other interventions involving lighting.

Stone E.L., Jones G. & Harris S. (2009) Street lighting disturbs commuting bats. *Current Biology*, 19, 1–5.
Stone E.L., Jones G. & Harris S. (2012) Conserving energy at a cost to biodiversity? Impacts of LED lighting on bats. *Global Change Biology*, 18, 2458–2465.

5.8 Replace or improve habitat for bats around roads

- We found no evidence for the effects of either replacing lost habitat or improving existing habitat around roads for bats.

Background
There is evidence that the effect of a road on bat diversity is reduced in better quality bat habitat (Berthinussen & Altringham 2012). Replacing lost habitat and improving habitat quality (for example by planting trees, hedges, woodland or creating wetlands) around roads may reduce the negative impact on bats. However, we found no evidence for the effects of habitat improvements around roads.

Berthinussen A. & Altringham J. (2012) The effect of a major road on bat activity and diversity. *Journal of Applied Ecology*, 49, 82–89.

6 Threat: Biological resource use

Key messages – Hunting

Introduce and enforce legislation to control hunting of bats
We found no evidence for the effects of introducing and enforcing legislation or regulations to control the hunting of bats.

Educate local communities about bats and hunting
We found no evidence for the effects of educating local communities to raise awareness about the importance of bats and the risks/negative impacts associated with hunting them.

Introduce sustainable harvesting of bats
We found no evidence for the effects of introducing methods of sustainable harvesting of bats.

Key messages – Guano harvesting

Introduce and enforce legislation to regulate the harvesting of bat guano
We found no evidence for the effects of introducing and enforcing legislation to regulate the harvesting of bat guano.

Introduce sustainable harvesting of bat guano
We found no evidence for the effects of introducing methods of sustainable harvesting of bat guano.

Key messages – Logging and wood harvesting

Use selective harvesting/reduced impact logging instead of clearcutting
Nine replicated, controlled, site comparison studies provide evidence for the effects of selective or reduced impact logging on bats with mixed results. One study in the USA found that bat activity was higher in selectively logged forest than in unharvested forest. One study in Italy caught fewer barbastelle bats in selectively logged forest than in unmanaged forest. Three studies in Brazil and two in Trinidad found no difference in bat abundance or species diversity between undisturbed control forest and selectively logged or reduced impact logged forest, but found differences in species composition. Two studies in Brazil found no effect of reduced impact logging on the activity of the majority of bat species, but mixed effects on the activity of four species.

Use shelterwood cutting instead of clearcutting
One site comparison study in North America found higher or equal activity of at least five bat species in shelterwood harvests compared to unharvested control sites. One replicated, site comparison study in Australia found Gould's long-eared bats selectively roosting in shelterwood harvests, but southern forest bats roosting more often in mature unlogged forest. A replicated, site comparison study in Italy found barbastelle bats favoured unmanaged woodland for roosting and used shelterwood harvested woodland in proportion to availability.

Thin trees within forests
Two replicated, site comparison studies (one paired) in North America found that bat activity was higher in thinned forest stands than in unthinned stands, and similar to that in mature forest. One replicated, site comparison study in North America found higher bat activity in thinned than in unthinned forest stands in one of the two years of the study. One replicated, site comparison study in Canada found the silver-haired bat more often in clearcut patches than unthinned forest, but found no difference in the activity of *Myotis* species.

Manage woodland or forest edges for bats
We found no evidence for the effects of managing forest or woodland edges as foraging habitat for bats.

Retain deadwood/snags within forests for roosting bats
We found no evidence for the effects of retaining large, old, dead or decaying trees in logged areas for roosting bats.

Replant native trees
We found no evidence for the effects of replanting native trees in logged areas.

Retain residual tree patches in logged areas
Two replicated, site comparison studies in Canada found no difference in bat activity between residual tree patch edges in clearcut blocks and edges of the remaining forest. One of the studies found higher activity of smaller bat species at residual tree patch edges than in the centre of open clearcut blocks. Bat activity was not compared to unlogged areas.

Incorporate forested corridors or buffers into logged areas
One replicated, site comparison study in Australia found no difference in the activity and number of bat species between riparian buffers in logged, regrowth or mature forest. One replicated, site comparison study in North America found higher bat activity along the edges of forested corridors than in corridor interiors or adjacent logged stands.

Three replicated, site comparison studies in Australia and North America found four bat species roosting in forested corridors and riparian buffers.

For all evidence relating to the use of **bat boxes/houses**, see 'Providing artificial roost structures for bats'.

Hunting

Background
Mostly fruit bat species, but also some insectivorous species, are hunted for bushmeat for both local and commercial consumption. Bats are also hunted for medicine or sport, and are culled as agricultural pests. There is evidence that the hunting of bats is having a significant impact on bat populations in the Old World tropics (Mickleburgh et al. 2009).

Mickleburgh S., Waylen K. & Racey P. (2009) Bats as bushmeat: a global review. *Oryx*, 43, 217–234.

6.1 Introduce and enforce legislation to control hunting of bats

- We found no evidence for the effects of introducing and enforcing legislation or regulations to control the hunting of bats.

Background
This intervention involves the introduction of legislation to protect bats from hunting. This may include measures such as hunting regulations, issue of hunting licences, prohibition of export, and the control of guns and ammunition. Subsequent enforcement of legislation is also important.

6.2 Educate local communities about bats and hunting

- We found no evidence for the effects of educating local communities to raise awareness about the importance of bats and the risks/negative impacts associated with hunting them.

Background
Education programmes that emphasize the role of bats in providing ecosystem services are being implemented in some countries, and may benefit local bat populations (Entwistle 2001, Mickleburgh *et al.* 2009). However, we found no evidence for the effects of educating local communities on bats.

Entwistle A. (2001) Community-based protection successful for the Pemba flying fox. *Oryx* 35, 355–356.
Mickleburgh S., Waylen K. & Racey P. (2009) Bats as bushmeat: a global review. *Oryx*, 43, 217–234.

6.3 Introduce sustainable harvesting of bats

- We found no evidence for the effects of introducing methods of sustainable harvesting of bats.

Background
Sustainable harvesting of the more abundant bat species and voluntary controls may help to reduce the decline of bat populations without threatening the traditions of local communities.

Guano harvesting

Background
Bat guano has a high concentration of nitrates and has been harvested from caves for centuries for a variety of uses. Modern day use is typically for fertilizer, both for commercial production and subsistence farming. Guano harvesting can cause serious disturbance to bat colonies resulting in arousal from hibernation, the abandonment of pups or total abandonment of caves as roosting sites. Proposed guidelines for the sustainable harvesting of bat guano in Southeast Asia have recently been drawn up by the EWCL (Emerging Wildlife Conservation Leaders) Bat Conservation Team (DiMiceli, 2012).

DiMiceli, C. (2012) Helping guano miners save bats – first international guidelines for sustainable guano harvests. *BATS*, 30, 8–9.

6.4 Introduce and enforce legislation to regulate harvesting of bat guano

- We found no evidence for the effects of introducing and enforcing legislation to regulate the harvesting of bat guano.

6.5 Introduce sustainable harvesting of bat guano

- We found no evidence for the effects of introducing methods of sustainable harvesting of bat guano.

Logging and wood harvesting

6.6 Use selective harvesting/reduced impact logging instead of clearcutting

- Nine replicated, controlled, site comparison studies[1–9] provide evidence for the effects of selective logging on bats with mixed results.

- One study in the USA[1] found higher bat activity in selectively logged forest than unharvested forest, and activity did not differ from that in high quality foraging habitat (Carolina bays).

- One study in Italy[9] caught fewer barbastelle bats in selectively logged forest than unmanaged forest, and fewer bats roosted in the logged forest.

- Three studies in Brazil[2, 4, 8] and two in Trinidad[6, 7] found no overall difference in bat abundance, species diversity or the distribution of bat species abundance patterns between undisturbed control forest and selectively logged forest. However, all studies found differences in species composition.

- Two studies in Brazil[3, 5] found no significant effect of reduced impact logging on the activity of the majority of bat species, but mixed effects on the activity of four species.

Background
Selective logging is the removal of selected trees within a forest based on criteria such as diameter, height or species. Remaining trees are left in the stand, as opposed to clearcutting where all trees are felled.

Reduced impact logging is a sustainable harvesting and management method that aims to minimize ecological disturbance. It involves selective logging as well as other practices such as directional tree felling, stream buffer zones, constructing roads, trails and landings to minimum widths and methods to extract timber with minimal damage. Studies on reduced impact logging have been included but the effects of selective logging cannot be separated from the other interventions used.

6.6 Use selective harvesting/reduced impact logging instead of clearcutting

In a replicated, controlled, site comparison study in June–August 1996 and 1997 in a forested site in west-central South Carolina, USA (1) bat activity was significantly lower in unharvested stands than in group selection harvested stands (average 1 vs. 7 bat passes/15 min respectively). There was no difference in bat activity between group selection harvested stands and Carolina bays (average 6 bat passes/15 min). Bat activity was found to be greater in gaps created by group selection harvesting and skidder trails than in forested areas (average 29 vs. 2 bat passes/detector/night). There was no difference in bat activity between gaps and skidder trails, large gaps (average 24 bat passes/detector/night) and small gaps (average 33 bat passes/detector/night), and at the centre (average 8 bat passes/detector/night) or edge of gaps (average 14 bat passes/detector/night). Unharvested stands were bottomland hardwoods in which no timber harvests had occurred in the last 70 years. Group selection harvested stands were bottomland hardwoods which had been harvested in December 1994 by mechanized methods to remove all commercial stems in 36 gaps of varying size (6 each of 0.02, 0.03, 0.06, 0.13, 0.26 and 0.5 ha). Carolina bays were wetlands surrounded by trees. Bat activity in stands was recorded from sunset with bat detectors along transects located in three sites of each stand type. Calls were recorded for one minute at 15 points every 10 m along 140 m transects in 5 four-day periods from June to August 1996. Within each stand, bat activity was recorded with bat detectors placed at the centre and edge of large gaps, in small gaps, along skidder trails and in the adjacent forest 40 m and 80 m from the cut edge, on 4 consecutive nights during 5 survey periods in June–August 1997. Eight bat species were caught at the sites with mist nets.

A replicated, controlled, site comparison study in 1999–2000 in a tropical lowland rainforest in western Pará, Brazil (2) found that the average abundance of 10 of the 15 most common bat species did not differ between control forest sites and reduced impact logging (RIL) sites. Three bat species were captured more in RIL sites than control sites (brown fruit-eating bat *Artibeus concolour*: average 0.8 vs. 0.3 bats/site respectively, great fruit-eating bat *Artibeus lituratus*: average 12 vs. 8 bats/site, Pallus's long-tongued bat *Glossophaga soricina*: average 1.1 vs. 0.4 bats/site). Two species were captured less in RIL sites than control sites (silky short-tailed bat *Carollia brevicauda*: average 0.8 vs. 1.8 bats/site respectively, Seba's short-tailed bat *Carollia perspicillata*: average 17 vs. 31 bats/site). Bat assemblages in RIL forest had increased diversity and decreased dominance, but the proportional abundance of the five most common species was similar in RIL (86%) and control forest sites (88%). Surveys were conducted in two undisturbed control forest blocks and two RIL blocks (all 100 ha). Within RIL blocks trees larger than 43 cm had been harvested at low intensity (18.7 m^3/ha). Timber harvest ceased 20 months before the study. Other methods were in place at RIL sites to reduce damage to the forest, such as directional felling of trees and a reduced number of landing sites and roads. Twelve sites within each block were sampled, with mist-netting at four sites per night from dusk to midnight or dawn in April–May 1999, November–December 1999 and April 2000.

A replicated, controlled, site comparison study in 1999–2000 in a tropical lowland rainforest in western Pará, Brazil (3) found no difference in the temporal activity of bats that were aerial insectivores, nectarivores and gleaning animalivores between control forest and reduced-impact logging (RIL) forest. Three of five fruit-eating bat species, all understorey foragers, were more active during the early evening in control forest than in RIL forest (great fruit-eating bat *Artibeus lituratus*: 46 vs. 77 bats captured at 1830 h in control and RIL forest respectively, silky short-tailed bat *Carollia brevicauda*: 40 vs. 11 bats, Seba's short-tailed bat *Carollia perspicillata*: 391 vs. 309 bats). Surveys were conducted in two undisturbed control forest blocks and two RIL blocks (all 100 ha). Within RIL blocks trees larger than 43 cm had been harvested at low intensity (18.7 m³/ha) and timber harvest ceased 20 months before the study. Other methods were in place at RIL sites to reduce damage to the forest, such as directional felling of trees and a reduced number of landing sites and roads. Twelve sites within each block were sampled, with mist netting at four sites per night from dusk to midnight or dawn in April–May 1999, November–December 1999 and April 2000.

In a replicated, controlled, site comparison study in 1999–2000 in a tropical lowland rainforest in western Pará, Brazil (4) 5 of the 17 most abundant bat species captured differed in abundance between reduced-impact logging (RIL) forest and unlogged forest. Four bat species were more abundant in RIL forest than unlogged forest (Heller's broad-nosed bat *Platyrrhinus helleri*: average 0.08 vs. 0.01 bats captured/night/site respectively, total 12 bats captured; gnome fruit-eating bat *Artibeus gnomus*: average 0.1 vs. 0.06 bats captured/night/site, total 25 bats; Pallas's long-tongued bat *Glossophaga soricina*: average 0.12 vs. 0.03 bats captured/night/site, total 23 bats; silky short-tailed bat *Carollia brevicauda*: average 0.21 vs. 0.08 bats captured/night/site, total 45 bats). One bat species (dwarf little fruit bat *Rhinophylla pumilio*) was less abundant in RIL forest than unlogged forest (average 0.05 vs. 0.13 bats captured/night/site respectively, total 35 bats captured). Surveys were conducted in two undisturbed control forest blocks and two RIL blocks (all 100 ha). Within RIL blocks trees larger than 43 cm had been harvested at low intensity (18.7 m³/ha) and timber harvest ceased 20 months before the study. Other methods were in place at RIL sites to reduce damage to the forest, such as directional felling of trees and a reduced number of landing sites and roads. Bats were surveyed for 4 nights at 96 sites evenly dispersed within the experimental blocks, with half of the sites in forest gaps and half in closed canopy forest. Average forest gaps were 219 m² in RIL sites and 286 m² in unlogged sites. Sampling was conducted during four time periods between June 1999 and April 2000, and each site was surveyed for one night during each time period. Bats were caught in mist nets at ground level open from 1800 to 0100 h. A total of 47 bat species were caught.

In a replicated, controlled, site comparison study in 1999–2000 in a tropical lowland rainforest in western Pará, Brazil (5) reduced impact logging (RIL) had no significant effect on the activity of six of the seven most commonly captured fruit-eating bat species. One species (the dark fruit-eating bat *Artibeus obscurus*) had significantly higher activity in forest gaps created by RIL than in natural gaps from falling trees in unlogged sites (e.g. relative abundance of 0.35 in RIL vs. 0.07 in unlogged sites, between 1800 and 1859 h). Surveys were conducted in two undisturbed control forest blocks and two RIL blocks (all 100 ha). Within RIL blocks trees larger than 43 cm had been harvested at low intensity (18.7 m³/ha) and timber harvest ceased 20 months before the study. Other methods were in place at RIL sites to reduce damage to the forest, such as directional felling of trees and a reduced number of landing sites and roads. Bats were surveyed for 4 nights at 96 sites evenly dispersed within the experimental blocks, with half of the sites in forest gaps and half in closed canopy forest. Average forest gaps were 219 m² in RIL sites and 286 m² in unlogged sites. Sampling was conducted during four time periods between June 1999 and April 2000, and each site was surveyed for one night during each time period. Bats were caught in mist nets at ground level open from 1800 to 0100 h. A total of 45 bat species were caught, dominated by fruit-eating species (1,169 of 1,468 captures).

A replicated, controlled, site comparison study in 2000–2002 in a lowland tropical forest in Victoria-Mayaro Forest Reserve, Trinidad (6) found that selective logging did not affect bat species diversity but did affect community structure. A total of 29 bat species were captured in unlogged forest and 34 in selectively logged forest. Fruit-eating bat species were significantly more abundant in logged forest (96–409 bats captured/site, 78–85% of captures) than unlogged

6.6 Use selective harvesting/reduced impact logging instead of clearcutting 38

forest (88–194 bats captured/site, 55–73% of captures). Gleaning species were less abundant in logged forest (11–24 bats captured/site, 3–9% of captures) than unlogged forest (34–37 bats captured/site, 14–21% of captures). The number of years since selective logging took place was positively correlated with the abundance and number of species of gleaning animalivores, and negatively correlated with the proportional abundance of the most common species, a generalist frugivore (Seba's short-tailed bat *Carollia perspicillata*). Surveys were conducted in two unlogged primary forest sites and five selectively logged sites using a periodic block system 33, 30, 21, 20 and 10 years previously. In the periodic block system, trees are selected and felled (4–8 trees/ha) in blocks of 150–300 ha based on several ecological criteria, including value to wildlife. The blocks are logged over one to two years and then closed for 30 years. In 2000–2002, bats were captured using mist nets (at ground level and in the forest subcanopy) and harp traps for six hours from sunset at five sampling points spaced 350–750 m apart twice at each site (once in the wet season and once in the dry season). A total of 38 bat species were caught.

A replicated, controlled, site comparison study in 2001–2002 in a lowland tropical forest in Victoria-Mayaro Forest Reserve, Trinidad (7) found a significant difference in the numbers of bats per guild between sites that were continuously logged, periodically logged or undisturbed. Fruit-eating bat species were more abundant in continuously logged forest (958 bats captured of 13 species, 87% of total captures) than periodically logged (352 bats captured of 9 species, 82% of total captures) or undisturbed forest (282 bats captured of 10 species, 66% of total captures). Gleaning animal-eating species were more abundant in undisturbed forest (71 bats captured of 9 species, 17% of total captures) than periodically logged (6 bats captured of 7 species, 6% of total captures) or continuously logged forest (52 bats captured of 8 species, 5% of total captures). The sample sizes of bats in other guilds were too small for comparisons to be made. The number of bat species or species captured per guild did not differ between undisturbed and logged forest. Continuously logged sites had been under open range management from 1954 to 1990 in which trees were cropped continuously with the only control being a girth limit for certain species. The canopy in open range forests is relatively open, allowing pioneer fruit plants to grow. In periodically logged sites, since the 1970s selected trees (4–8 trees/ha) in blocks of 150–300 ha were felled based on several ecological criteria including value to wildlife. The blocks are logged over 1–2 years and then closed for 30 years. Undisturbed sites were protected primary forest. Two sites were surveyed in each of the 3 habitats: primary undisturbed forest, forest under period block management (closed 10–20 years ago) and forest under open range management (closed 10 years ago). Bats were captured using mist nets (at ground level and in the forest subcanopy) and harp traps for 6 hours after sunset at 5 sampling points spaced 350–750 m apart twice at each site. A total of 1,959 bats were caught representing 38 different species.

A replicated, controlled, site comparison study in September–December 2002 in a tropical lowland rainforest in southern Pará, Brazil (8) found that overall bat abundance and species richness among taxonomic groups did not differ significantly between selectively logged and unlogged forest. One bat species, the tent-making bat *Uroderma bilobatum*, was caught significantly more often in selectively logged forest (56 bats caught in total vs. 11 bats in unlogged forest). One bat species, Parnell's mustached bat *Pteronotus parnellii*, was caught only in unlogged forest (16 bats caught in total). Capture success of bats was similar in understorey nets in selectively logged forest (42 bats/100 mist net hours) and unlogged forest (43 bats/100 mist net hours), but was 1.5 times higher in canopy nets in selectively logged forest (74 bats/100 mist net hours) than unlogged forest (55 bats/100 mist net hours). Understorey species composition differed between sites with Stenodermatine species associated more with logged sites, and Phyllostominae species associated with unlogged sites. Surveys were completed in 5 1 ha study grids in an area selectively logged for mahogany in 1992 (1–4 trees/ha), and 5 1 ha study grids in an 8,000 ha unlogged forest reserve 4 km away. Bats were caught in mist nets (from 1800 to 0000 hours) in the understorey and canopy. Each grid was sampled two or three times, usually for two consecutive nights, with two week intervals between sampling sessions in the same grid. A total of 49 bat species were caught.

In a replicated, controlled, site comparison study in June–August 2001–2008 in two beech *Fagus sylvatica* forests in central Italy (9) barbastelle bats *Barbastella barbastellus* were caught more often in unmanaged forest (average 0.19 bats/capture effort/session) than selectively logged forest (average 0.07 bats/capture effort/session). Females were caught three times less often in the selectively logged forest (average 0.05 bats/capture effort/session) than in the unmanaged forest (average 0.16 bats/capture effort/session). Fourteen radiotracked bats caught in the logged forest were followed to 16 roosts, 5 of which were in dead trees in the unmanaged forest. Twenty-five radiotracked bats caught in the unmanaged forest were followed to 38 roosts in the same forest, and none moved to the logged area. The logged forest consisted of young beech stands periodically logged by plot rotation and selective logging (logged trees selected by a forestry technician based on diameter, position etc.). The logged forest had a greater canopy closure, fewer dead trees, a smaller tree diameter and trees with fewer cavities than the unmanaged forest. Bats were sampled using mist nets (for four hours from dusk) near cattle troughs used by bats for drinking, and capture frequency was compared between sites. Moth numbers were sampled with a light trap on two consecutive nights in July 2008 in both forests at three capture sites, and did not differ between unmanaged and logged forest.

(1) Menzel M.A., Carter T.C., Menzel J.M., Ford W.M. & Chapman B.R. (2002) Effects of group selection silviculture in bottomland hardwoods on the spatial activity patterns of bats. *Forest Ecology and Management*, 162, 209–218.
(2) Castro-Arellano I., Presley S.J., Saldanha L. N., Willig M.G. & Wunderle J.M. Jr. (2007) Effects of reduced-impact logging on bat biodiversity in terra firme forest of lowland Amazonia. *Biological Conservation*, 138, 269–285.
(3) Castro-Arellano I., Presley S.J., Willig M.R., Wunderle J.M. & Saldanha L.N. (2009) Reduced-impact logging and temporal activity of understorey bats in lowland Amazonia. *Biological Conservation*, 142, 2131–2139.
(4) Presley S.J., Willig M.R., Wunderle J.M. Jr. & Saldanha L.N. (2008) Effects of reduced-impact logging and forest physiognomy on bat populations of lowland Amazonian forest. *Journal of Applied Ecology*, 45, 14–25.
(5) Presley S.J., Willig M.R., Saldanha L.N., Wunderle J.M. Jr. & Castro-Arellano I. (2009) Reduced-impact logging has little effect on temporal activity of fruit-eating bats (Chiroptera) in lowland Amazonia. *Biotropica*, 41, 369–378.
(6) Clarke F.M., Rostant L.V. & Racey, P.A. (2005) Life after logging: post-logging recovery of a neotropical bat community. *Journal of Applied Ecology*, 42, 409–420.
(7) Clarke F.M., Pio D.V. & Racey P.A. (2005) A comparison of logging systems and bat diversity in the Neotropics. *Conservation Biology*, 19, 1194–1204.
(8) Peters S.L., Malcolm J.R. & Zimmerman B.L. (2006) Effects of selective logging on bat communities in the southeastern Amazon. *Conservation Biology*, 20, 1410–1421.
(9) Russo D., Cistrone L., Garonna, A.P. & Jones, G. (2010) Reconsidering the importance of harvested forests for the conservation of tree-dwelling bats. *Biodiversity and Conservation*, 9, 2501–2515.

6.7 Use shelterwood cutting instead of clearcutting

- One site comparison study in North America[2] found higher or equal activity of at least five bat species in shelterwood harvests than unharvested control sites.

- One replicated, site comparison study in Australia[3] found Gould's long-eared bats selectively roosting in shelterwood harvests, but southern forest bats roosting more often in

6.7 Use shelterwood cutting instead of clearcutting

mature unlogged forest. A replicated site comparison study in Italy[1] found barbastelle bats favoured unmanaged woodland for roosting and used shelterwood harvested woodland in proportion to availability.

Background
There are several different shelterwood systems. The basic process is the selective removal of overstorey trees to allow enough light through to the forest floor to create new, even-aged stands below. The remaining mature overstorey trees provide seeds for regeneration and create shelter for the younger trees. Harvesting is done in a series of cuts, and may also involve thinning of the lower forest canopies.

In a replicated, site comparison study in July and August 2001–2002 in beech *Fagus sylvatica* woodlands in central Italy (1) barbastelle bats *Barbastella barbastellus* were found to use shelterwood harvested woodland for roosting in proportion to availability, whereas unmanaged woodland was positively selected, and pasture interspersed with woodland was avoided (4 roosts in 69 ha of shelterwood, 19 roosts in 250 ha of unmanaged woodland and 10 roosts in 381 ha of pasture). Shelterwood harvested stands were selectively harvested (understory trees only) to reduce competition among trees in 1997–2002. Unmanaged woodland had not been logged for at least 40 years. Bats were caught using mist nets near cattle troughs used by bats for drinking, and fitted with radio transmitters. Twenty-five adult barbastelle bats were followed to 33 roosts where roost characteristics were recorded and compared to random trees. Most roost trees were dead, and were significantly taller, with a larger diameter and more cavities, than random trees.

A site comparison study in May–September 2006 in two oak-hickory forests *Carya* spp. in southern Ohio, USA (2) found activity of three bat species (eastern red bat *Lasiurus borealis*, big brown bat *Eptesicus fuscus* and silver-haired bat *Lasionycteris noctivagans*) in shelterwood harvested sites but not in unharvested control sites (50% eastern shelterwood: eastern red bat average 1.1 passes/site, big brown bat and silver-haired bat combined average 0.6 bat passes/site; 70% shelterwood: red bat average 0.6 passes/site, big brown bat and silver-haired bat combined average 0.8 bat passes/site; control sites: 0 bat passes). There was no difference in the activity of *Myotis* spp. and tri-colored bats *Perimyotis subflavus* between harvested sites (50% shelterwood: combined average 0.2 passes/night, 70% shelterwood: 0.24 passes/night) and unharvested control sites (combined average 0.07 passes/night). There was no difference in activity for any of the species between two harvesting retention levels (50% and 70% full stocking). Bat activity declined as structural volume in the understorey to the mid-canopy of the shelterwood harvests increased. The average number of bat passes decreased by 50% at volumes exceeding 17 m^3/ha within 3–6 m of the forest floor. Estimated use by eastern red bats decreased by 50% when volumes within 0–12 m exceeded 1,750 m^3/ha, and estimated use by big brown bats and silver-haired bats was highest when volumes within 3–6 m exceeded 63 m^3/ha. In each study area 2 replicates of a control and 2 shelterwood harvests (50% and 70% stocking levels) of 10 ha were surveyed. In the shelterwood harvests a combined crown thinning and low thinning, favouring dominant and co-dominant oak, was done from March 2005 to June 2006 to reduce the stocking level. A random selection of two of the study sites was sampled per night with each site sampled for six to eight nights. The vegetation was sampled in eight plots within each site. Bat activity was recorded for 3 hours from 30 minutes before sunset in the same 8 plots using bat detectors. Mist netting was conducted during June–August 2006 for one to two nights per week.

In a replicated, site comparison study in February–March 2009 in jarrah *Eucalyptus marginata* forests in south-western Australia (3) tracked Gould's long-eared bats *Nyctophilus gouldi* were found to selectively roost in remnant trees in shelterwood harvests (10 bats, 37%). The remainder of tracked Gould's long-eared bats roosted in gap release systems (1 bat, 3%), mature forest (8 bats, 30%) and riparian buffers (8 bats, 30%). Only one southern forest bat *Vespadelus regulus* tracked during the study roosted in shelterwoods. Southern forest bats roosted more in mature unlogged forest (15 bats, 71%) and riparian buffers (five bats, 24%). Bats changed roosts every one to two days but showed fidelity to a general area. Shelterwoods had retention levels of 40–60%. Gap release systems involved the removal of 95% of the mature overstory. Riparian buffers and mature forest areas had been undisturbed for more than 30 years with only light selective logging prior to this. Bats (mostly females) were caught at two different water holes with different logging histories and fitted with radio transmitters. Ten southern forest bats were successfully tracked to 21 different roosts for an average of 6 days. Eleven Gould's long-eared bats were successfully tracked to 27 roosts for an average of 4 days. Characteristics of roost trees were recorded and compared to random trees, and roost availability in the surrounding forest landscape was estimated.

(1) Russo D., Cistrone L., Jones G. & Mazzoleni S. (2004) Roost selection by barbastelle bats (*Barbastella barbastellus*, Chiroptera: Vespertilionidae) in beech woodlands of central Italy: Consequences for conservation. *Biological Conservation*, 117, 73–81.
(2) Titchenell M.A., Williams R.A., Gehrt S.D. (2011). Bat response to shelterwood harvests and forest structure in oak-hickory forests. *Forest Ecology and Management*, 262, 980–988.
(3) Webala P.W, Craig M.D., Law B.S., Wayne A.F. & Bradley J.S. (2010) Roost site selection by southern forest bat *Vespadelus regulus* and Gould's long-eared bat *Nyctophilus gouldi* in logged jarrah forests; south-western Australia. *Forest Ecology and Management*, 260, 1780–1790.

6.8 Thin trees within forest

- Two replicated, site comparison studies (one paired) in North America[1, 2] found that bat activity was higher in thinned forest stands than in unthinned stands, and similar to that in mature forest.

- One replicated, site comparison study in North America[4] found higher bat activity in thinned than in unthinned forest stands in one of the two years of the study.

- One replicated, site comparison study in Canada[3] found the silver-haired bat more often in clearcut patches than unthinned forest, but found no difference in the activity of *Myotis* species.

Background
Thinning is a forestry practice that involves the selective removal of trees to reduce tree density and improve the growth rate and health of remaining trees. Thinning has been done historically to maximize timber production, but may have ecological benefits. The retention of large old trees may provide roost sites for bats, and opening up the canopy may provide favourable foraging habitats.

A replicated, site comparison study in summer 1993 and 1994 in intensively managed forests in the Cascade mountains, USA (1) recorded higher levels of bat activity in clearcut stands than in pre-commercially thinned stands, young unthinned stands or mature stands (average

7.5 bat passes/night in clearcut vs. 2 bat passes/night in pre-commercially thinned, 0 bat passes/night in young unthinned vs. 4 bat passes/night in mature stands). Six replicates of stands in four post-harvest stages were sampled. Clearcut stands (2–3 years post-harvest) had seedlings of Douglas fir *Pseudotsuga menziesii* 1–2 m high. Pre-commercially thinned stands (12–20 years post-harvest) had 10–13 year old Douglas fir stands with light reaching the ground between trees. Young unthinned stands (30–40 years old) had a high tree density with a range of tree diameters. Mature stands (51–62 years old) were commercially thinned stands dominated by Douglas fir or western hemlock *Tsuga heterophylla*. Bat activity was monitored with bat detectors at each site for six full nights from July–September in each year. Four bat species were identified in the study, and an unknown number of *Myotis* spp. (seven are known to be present in the area).

In a replicated, paired, site comparison study in summer 1994 and 1995 in 11 pairs of forest stands in the Oregon Coast range, USA (2) total bat activity (of at least 9 species) was 1.6 times higher in thinned than unthinned stands (average 10 vs. 6 bat passes/night). There was no significant difference in total bat activity between thinned and old growth forest (average 13 bat passes/night). Stands were 50–100 years old with sections that had been thinned in 1971–1985. Old growth forest was over 200 years old with minimal human disturbance. All stands were dominated by Douglas fir *Pseudotsuga menziesii*. Pairs of thinned and unthinned stands of 10 ha were matched for elevation, slope and aspect. Bat activity was recorded using bat detectors at random locations in pairs of stands and nearby old growth forest simultaneously for two full consecutive nights on four occasions. Seven sites were sampled in 2005 and four in 1995, with two of the sites surveyed in both years. Characteristics of vegetation and habitat structure were also measured in 50 m radius plots around each survey point. Old growth stands had large diameter trees and abundant snags. Thinned and unthinned stands had few large trees or snags.

A replicated, controlled, site comparison study in 1999–2000 in a mixed wood forest of Alberta, Canada (3) found that habitat selection by bats in patches of clearcut, thinned and unthinned forest varied for different species. The presence and activity of *Myotis* species (including little brown bats *Myotis lucifugus* and northern long-eared bats *Myotis septentrionalis*) did not differ significantly between patch types. The silver-haired bat *Lasionycteris noctivagans* was present more often in clearcut patches than thinned or unthinned patches (absent on all but one night), and activity levels were significantly higher in clearcut patches (e.g. -0.4 log bat passes/hour at the centre of clearcut deciduous patches) than thinned patches in all types of forest (e.g. -0.6 log passes/hour at the centre of both 20% and 50% thinned deciduous patches). Results are given as least-squares means. Bat activity was measured in forest patches of four tree densities (0% clearcut, 20% and 50% thinned, 100% unthinned) in three forest types (deciduous, coniferous and mixed wood). Harvesting was done in the study area in winter 1998–1999 to create 10 ha experimental forest patches for each treatment of tree density (three replicates per treatment in each forest type). All forest types were over 50 years old. Each patch was surrounded by a buffer of intact forest. Bat activity was recorded with bat detectors (from 30 minutes after sunset until 150 minutes or 1 hour after sunset) at the centre and edge of each patch in June–July 1999 and June–August 2000. Over the 2 years, each location within each patch of each of the 3 replicates was sampled for 11–14 nights in total.

A replicated, controlled, site comparison study in May–August 2001 and 2002 in an experimental pine forest in South Carolina, USA (4) found that bat activity was higher in thinned stands than in unthinned control stands but this difference was only statistically significant in 2001 (average 8.2 vs. 1.3 bat passes/night respectively in 2001, average 2.1 vs.

0.4 bat passes/night in 2002). In both years of the study, overall bat activity was intermediate in burned stands (average 2.1 bat passes/night in 2001, 2 bat passes/night in 2002) and thinned/burned stands (average 3.4 bat passes/night in 2001, 1.9 bat passes/night in 2002), but not significantly different from control stands without any treatment. Of the three most frequently recorded species, this pattern was consistent for big brown bats *Eptesicus fuscus* and eastern red bats *Lasiurus borealis*, but not for eastern pipistrelles *Perimyotis subflavus*, which did not vary between stand types in either year. Twelve 14 ha stands were selected with 3 replicates of 4 treatment types: prescribed burning (burned in April 2001 with strip head fire and flanking fires), thinning to 18 m^2/ha (in winter 2000–2001), thinning to 18 m^2/ha followed by prescribed burning (burned in spring 2002 with strip head fires) and a control with no treatment. Bat activity was sampled from sunset to sunrise with two bat detectors at random points in each stand for two nights per month from May–August in both years.

(1) Erickson, J.L. & West S.D. (1996) Managed forests in the western Cascades: the effects of seral stage on bat habitat use patterns. In Barclay, R.M.R. and Brigham, R.M. (eds.) *Bats and Forests Symposium*. British Columbia Ministry of Forests, Victoria, Canada. pp. 215–227.
(2) Humes M.L., Hayes J.P. & Collopy M.W. (1999) Bat activity in thinned, unthinned, and old-growth forests in western Oregon. *The Journal of Wildlife Management*, 63, 553–561.
(3) Patriquin K.J. & Barclay R.M.R. (2003) Foraging by bats in cleared, thinned and unharvested boreal forest. *Journal of Applied Ecology*, 40, 646–657.
(4) Loeb S.C. & Waldrop T.A. (2008) Bat activity in relation to fire and fire surrogate treatments in southern pine stands. *Forest Ecology and Management*, 255, 3185–3192.

6.9 Manage forest or woodland edges for bats

- We found no evidence for the effects of managing forest or woodland edges as foraging habitat for bats.

Background
Edge habitats are important for foraging bats. A study in North America found significantly higher activity of six bat species along forest edges than forest interiors in both unmanaged and thinned forest (Morris *et al.* 2010).

Morris A.D., Miller D.A. & Kalcounis-Rueppell M.C. (2010) Use of forest edges by bats in a managed pine forest landscape. *The Journal of Wildlife Management*, 74, 26–34.

6.10 Retain deadwood/snags within forests for roosting bats

- We found no evidence for the effects of retaining large, old, dead or decaying trees in logged areas for roosting bats.

Background
Tree-roosting bats have been found to prefer old growth forest stands (Thomas 1988), and select trees to roost in that are typically tall, large in diameter and dead or decaying ('snags') often with hollows, cracks and cavities or exfoliating bark (Campbell at al. 1996, Crampton & Barclay 1998, Rabe *et al.* 1998). Zielinski and Gellman (1999)

found that bats used hollows in old growth remnant redwood stands more than those in intact forest.

See 'Threat: Residential and commercial development – Provide foraging habitat in urban areas' for one study that uses snag recruitment alongside other practices for forest restoration.

Campbell L.A., Hallett J.G. & O'Connell M.A. (1996) Conservation of bats in managed forests: use of roosts by *Lasionycteris noctivagans*. *Journal of Mammalogy*, 77, 976–984.
Crampton L.H. & Barclay R.M.R. (1998) Selection of roosting and foraging habitat by bats in different-aged Aspen mixed wood stands. *Conservation Biology*, 12, 1347–1358.
Rabe M.J., Morrell T.E., Green H., de Vos J.C. Jr. & Miller C.R. (1998) Characteristics of ponderosa pine snag roosts used by reproductive bats in northern Arizona. *The Journal of Wildlife Management*, 62, 612–621.
Thomas D.W. (1988) The distribution of bats in different ages of Douglas-fir forests. *The Journal of Wildlife Management*, 52, 619–626.
Zielinski W.J. & Gellman S.T. (1999) Bat use of remnant old-growth redwood stands. *Conservation Biology*, 13, 160–167.

6.11 Replant native trees

- We found no evidence for the effects of replanting native trees in logged areas.

See 'Threat: Agriculture – Land use change – Retain or plant trees to replace foraging habitat for bats' for evidence relating to the replanting of native trees in areas cleared for agriculture.

6.12 Retain residual tree patches in logged areas

- Two replicated, site comparison studies in Canada[1, 2] found no difference in bat activity along the edge of residual tree patches in clearcut blocks and the edge of the remaining forest. One study found higher activity of smaller bat species along the edge of residual tree patches than in the centre of open clearcuts[1]. Neither study compared bat activity in the residual tree patches to that in tree patches or forests in unlogged areas.

Background
Logging by clearcutting results in large open, cleared areas in forests (clearcut blocks). Residual tree patches may be left uncut within these areas.

A replicated, site comparison study from June to July 2000 in experimental forest in Alberta, Canada (1) found that more bat passes of smaller bat species (calls detected at 45 kHz) were recorded at forest edges and residual patch edges than in the centre of open clearcut blocks (average five and four bat passes/hour at forest and residual patch edges respectively vs. two bat passes/hour at the centre of open clearcut blocks). No significant difference was found between the two edge types. The number of passes of larger bat species (calls detected at 25 kHz) and the foraging activity of small and large bats was not significantly different between any of the locations. Bat activity was sampled at dusk for 27 nights using bat detectors along the edge of residual patches, along the forest edge of clearcut blocks and in the centre of clearcut blocks. Three clearcut blocks were sampled. Each location within a clearcut block was sampled two or three times in a randomized order. Small bats were *Myotis* spp. and large bats were either big brown bats *Eptesicus fuscus*, hoary bats *Lasiurus cinereus*, or silver-haired bats *Lasionycteris noctivagans*. Bat activity in residual tree patches was not compared to tree patches or forest in unlogged areas.

A small, replicated, site comparison study in July and August 2000 in logged forests in central British Columbia, Canada (2) found no significant difference in bat activity between clearcut forest edges and residual patch edges (49 vs. 110 total bat passes respectively). Residual patch edges at an intermediate distance from the clearcut edge of the forest had the highest bat activity (49 total bat passes vs. 3 and 8 total bat passes at residual patch edges closest to or furthest from clearcut forest edge respectively). Bat activity was recorded with bat detectors simultaneously at patch edges and clearcut edges from dusk until dawn for one night. Six residual patches were sampled in six different clearcut blocks with varying distances from the clearcut edge. All residual patches were 0.5–2 ha in size located in clearcuts of less than 5 years old ranging in size from 105–180 ha. The area is dominated by pine and spruce interspersed with aspen. Five bat species were known to occur in the area and all activity was pooled. Bat activity in residual tree patches was not compared to tree patches or forest in unlogged areas.

(1) Hogberg L.K., Patriquin K.J. & Barclay R.M.R. (2002) Use by bats of patches of residual trees in logged areas of the boreal forest. *American Midland Naturalist*, 148, 282–288.
(2) Swystun M.B, Syllakis J.M & Brigham R.M. (2001) The influence of residual tree patch isolation on habitat use by bats in central British Columbia. *Acta Chiropterologica*, 3, 197–201.

6.13 Incorporate forested corridors or buffers into logged areas

- One replicated, site comparison study in Australia[1] found no difference in bat activity and the number of bat species between riparian buffers in logged, regrowth and mature forest.

- One replicated, site comparison study in North America[4] found higher bat activity along the edges of forested corridors than in corridor interiors or adjacent logged stands.

- Three replicated, site comparison studies in Australia[5] and North America[2,3] found four bat species roosting in forested corridors and riparian buffers. One species roosted more often in forested corridors than unlogged forest[2], another species roosted equally in both[5], and two species roosted more often in unlogged forest[3,5].

Background
This intervention involves retaining corridors of unlogged mature forest, or leaving unlogged buffers around streams and rivers. This may provide foraging and roosting opportunities for bats, and maintain connectivity in disturbed landscapes.

A replicated, site comparison study from January to April 2003 in 60 forest sites near Coff's Harbour, New South Wales, Australia (1) found no significant difference in bat activity and the number of bat species in riparian buffers in logged forest, regrowth forest and mature forest (average 1.9 bat passes/hour and 0.32 species/hour in logged vs. 1.5 bat passes/hour and 0.3 species/hour in regrowth vs. 1.4 bat passes/hour and 0.36 species/hour in mature). One species, the eastern forest bat *Vespadelus pumilus*, was 2.7 times more active in riparian buffers in logged forest than in mature forest. Higher bat activity was found along larger streams than small streams, but this pattern was unaffected by logging history. Bat activity was found to be higher along forest tracks on upper slopes than riparian buffers with small

6.13 Incorporate forested corridors or buffers into logged areas

streams, but levels were similar along buffers with large streams. Five replicates of four sizes of stream were sampled for three different logging treatments (logged, regrowth and mature forest). Logged forest had been thinned and/or selectively logged in the last six years, regrowth forest had been logged 15–30 years ago, and mature forest had been undisturbed for more than 50 years. Bat activity was recorded using six bat detectors at six sites per night. Twenty-eight species were recorded in the study. The 60 riparian buffers were also paired with forest tracks running parallel on upper slopes and sampled simultaneously. Riparian buffers (logging exclusion zones along streams) varied in size (10–50 m minimum width) and were wider for larger streams.

A replicated, site comparison study in May–August 2003–2006 in intensively managed pine woodland in southern South Carolina, USA (2) found 61% of male and 63% of female Seminole bats *Lasiurus seminolus* roosting in forested corridors (25 and 31 roosts respectively). Thirty-four percent (14 roosts) of male and 29% (14 roosts) of female bats roosted in logged mid-rotation stands, and the remainder roosted in mature forest. Distance to the nearest forest corridor was negatively related with roost site selection in male and non-reproductive female bats. Distance to nearest edge and distance to mature pine stands (negatively related with roost site selection) were more important variables for reproductive females. Bats were caught from May to August in 2003–2006 at nine random ponds in open habitat throughout the study area and radio transmitters were attached. Twenty-seven adult Seminole bats (10 males and 17 females) were tracked to 90 day roosts. Characteristics of roost trees were recorded and compared to random trees. The 41,365 ha study area is intensively managed for the production of loblolly pine *Pinus taeda* with even-aged stands in various successional stages. Stands are clearcut at 20–25 years of age and are managed by commercial thinning, fire, mechanical and chemical treatments. Mid-rotation stands were 12–22 years old with almost complete canopy closure. Forested corridors (100–200 m wide) of mature pine and/or mixed hardwood cover 11% of the study area. Bats roosted in the canopy of tall, large, live pine trees.

A replicated, site comparison study in May–August 2003–2006 in intensively managed pine woodland in southern South Carolina, USA (3) found 39% of male and 18% of female evening bats *Nycticeius humeralis* roosting in forested corridors (12 and 8 roosts respectively). Nineteen percent (6 roosts) of male and 21% (9 roosts) of female bats roosted in logged mid-rotation stands, and the remainder roosted in mature forest. Distance to the nearest forest corridor was negatively related with roost site selection in male bats but not lactating females. Bats were caught from May to August in 2003–2006 at nine random ponds in open habitat throughout the study area and radio transmitters were attached. Fifty-three adult evening bats (26 male and 27 female) were tracked to 75 day roosts. Characteristics of roost trees were recorded and compared to random trees. The 41,365 ha study area is intensively managed for the production of loblolly pine *Pinus taeda* with even-aged stands in various successional stages. Stands are clearcut at 20–25 years of age and are managed by commercial thinning, fire, mechanical and chemical treatments. Mid-rotation stands were 12–22 years old with almost complete canopy closure. Forested corridors (100–200 m wide) of mature pine and/or mixed hardwood cover 11% of the study area. Bats roosted in nine tree species, in tree cavities, fork topped pines, live trees and exfoliating bark.

A replicated, site comparison study from June to August in 2004 and 2005 in intensively managed pine woodland in southern South Carolina, USA (4) recorded higher activity of 6 bat species along forested corridor edges than in corridor interiors or adjacent logged stands (54 call sequences/detector/night along corridor edges vs. 7 in corridor interiors vs. 12 in logged stands). The occupancy of three bat species (the big brown bat *Eptesicus fuscus*, the

Seminole bat *Lasiurus seminolus*, and the eastern pipistrelle *Perimyotis subflavus*) was higher when forested corridors were adjacent to roads. Bat activity was simultaneously recorded using bat detectors in 32 pairs of forested corridors (designed to either protect wet areas or enhance biodiversity) and adjacent logged pine stands from sunset to sunrise for 2 consecutive nights. The study area was 41,365 ha of loblolly pine *Pinus taeda* stands in various successional stages (clearcut at 20–25 years of age and managed by commercial thinning, fire, mechanical and chemical treatments) with a system of 100–200 m corridors of mature forest (11% of the total study area). Bat activity was not compared to that in unlogged forest.

In a replicated, site comparison study in February–March 2009 in jarrah *Eucalyptus marginata* forests in south-western Australia (5), both the southern forest bat *Vespadelus regulus* and Gould's long-eared bat *Nyctophilus gouldi* roosted in riparian unlogged buffers. The southern forest bat roosted more in mature unlogged forest (71%, 15 roosts) than in riparian buffers (24%, 5 roosts). One roost was found in a shelterwood logged stand (5%). Gould's long-eared bat roosted equally in mature unlogged forest and riparian buffers (both 30%, 8 roosts), and also in remnant trees in logged areas (36%, 10 roosts in shelterwood logged stands; 4%, 1 roost in gap release stands). Bats changed roosts every one to two days but showed fidelity to a general area. Riparian buffers were unlogged zones around drainage lines and streams. Both riparian buffers and mature forest areas had been undisturbed for more than 30 years with only light selective logging prior to this. Gap release stands had 95% of the mature canopy removed with 10–13 habitat trees (with hollows for fauna) retained per hectare. Shelterwood stands retained 40–60% basal area after logging with variable gaps of less than 10 ha. Bats (mostly females) were caught at two different water holes with different logging histories and fitted with radio transmitters. Ten southern forest bats were successfully tracked to 21 different roosts for an average of 6 days. Eleven Gould's long-eared bats were successfully tracked to 27 roosts for an average of 4 days. Characteristics of roost trees were recorded and compared to random trees, and roost availability in the surrounding forest landscape was estimated. Riparian buffers and mature forest contained a higher density of older large diameter trees with hollows.

(1) Lloyd A., Law B. & Goldingay R. (2006) Bat activity on riparian zones and upper slopes in Australian timber production forests and the effectiveness of riparian buffers. *Biological Conservation*, 129, 207–220.
(2) Hein C.D., Castleberry S.B. & Miller K.V. (2008) Sex-specific summer roost-site selection by Seminole bats in response to landscape-level forest management. *Journal of Mammalogy*, 89, 964–972.
(3) Hein C.D., Miller K.V. & Castleberry S.B. (2009) Evening bat summer roost-site selection on a managed pine landscape. *The Journal of Wildlife Management*, 73, 511–517.
(4) Hein C.D., Castleberry S.B. & Miller K.V. (2009) Site-occupancy of bats in relation to forested corridors. *Forest Ecology and Management*, 257, 1200–1207.
(5) Webala P.W, Craig M.D., Law B.S., Wayne A.F. & Bradley J.S. (2010) Roost site selection by southern forest bat *Vespadelus regulus* and Gould's long-eared bat *Nyctophilus gouldi* in logged jarrah forests; south-western Australia. *Forest Ecology and Management*, 260, 1780–1790.

7 Threat: Human disturbance – caving and tourism

Key messages

Use cave gates to restrict public access
Ten studies in Europe, North America and Australia provide evidence for the effects of cave gating on bats, with mixed results. Four of the studies (one replicated) found more or equal numbers of bats in underground systems after gating. Two of the studies (one replicated) found reduced bat populations or incidences of cave abandonment after gating. Five studies (two replicated) provide evidence for changes in flight behaviour at cave gates.

Maintain microclimate at underground hibernation/roost sites
We found no evidence for interventions that maintain the microclimate at roost sites.

Impose restrictions on cave visits
Two before-and-after studies from Canada and Turkey found that bat populations within caves increased after restrictions on cave visitors were imposed.

Educate the public to reduce disturbance to hibernating bats
We found no evidence for the effects of educating the public about bats to reduce disturbance at hibernations sites such as caves.

Legally protect bat hibernation sites
We found no evidence for the effects of legally protecting important bat hibernation sites that may be subject to human disturbance.

Provide artificial hibernacula for bats to replace disturbed sites
We found no evidence for the effects of providing artificial hibernacula for bats to replace hibernation sites lost due to human disturbance.

7.1 Use cave gates to restrict public access

- Ten studies in Europe, North America, Canada and Australia provide evidence for the effects of cave gating on bats, with mixed results.

- Four studies (one of which was a replicated, before-and-after trial) found more or equal numbers of bats in underground systems after gates were installed to restrict public access[1, 2, 5, 10].

- Two studies (both before-and-after, one replicated) found mines were abandoned or had reduced bat populations after gating of entrances[7, 8].

- Five studies (including two replicated, controlled, before-and-after and site comparison trials) provide evidence for changes in bat flight behaviour when cave gates are installed[3, 4, 6, 7, 9], and an effect of cave gate design[6, 7, 9].

49 Threat: Human disturbance – caving and tourism

Background

Recreational users of caves can disturb both hibernating and nursing colonies of bats causing abandonment of young or arousal from hibernation. Gates have been installed at cave entrances to restrict public access and reduce human disturbance. However, cave gating can also impede access by bats and early attempts from the 1950s to the 1970s often resulted in roost abandonment (Tuttle 1977). Gates that are more 'bat-friendly' have since been designed. Seven studies provide evidence for the effects of cave gates on bat populations. Five studies are also included which provide evidence for changes in flight behaviour and the effect of gate design on bats.

Tuttle M.D. (1977) Gating as a means of protecting cave dwelling bats. In Aley, T. & Rhodes, D. (eds.) *1976 National Cave Management Symposium Proceedings*, Speleobooks, Albuquerque, USA. pp. 77–82.

An unreplicated, site comparison study from 1976 to 1977 in two caves in Indiana, USA (1) found that Indiana bats *Myotis sodalis* hibernating within a cave modified with a stone wall and gate constructed at the entrance entered hibernation at a 5% higher body mass and lost 42% more body mass than bats in an unmodified cave 4 km away. The stone wall and gate in the modified cave restricted the cave opening by 62% reducing airflow and resulting in average winter temperatures 5°C higher than in the unmodified cave. Cave temperatures were measured near to hibernation sites every other week, and bats were counted and weighed in early winter (October–November 1976) and late winter (March 1977). In 1977, the stone wall was removed and replaced with steel bars. A biannual census until 1991 reported a subsequent increase in the population of Indiana bats in the cave from 2,000 to 13,000 bats.

In a controlled, before-and-after study from 1976 to 1984 in 3 abandoned underground war bunkers in the Netherlands (2) the number of hibernating bats was found to increase one and a half years after human access had been restricted by installing grilles or sealing entrances (from 35 to 115 bats in bunker 1, 35 to 50 bats in bunker 2, and 15 to 30 bats in bunker 3). Counts of bats were dominated by Daubenton's bats *Myotis daubentonii*, and small populations of pond bats *Myotis dasycneme* were also found to increase. A few individuals of three other species were also recorded. Bat numbers at a comparable bunker with no protection measures in place remained constant. Annual winter counts were conducted from 1976 or 1978 until 1984. Bunker entrances were either sealed completely or grilles of vertical bars were installed in 1977 or 1980. Sand and debris were also removed from one of the bunkers. The individual effects of each protection measure are not known.

A small replicated study in summer 1985 at two caves in Alabama and West Virginia, USA (3) found that Townsend's big-eared bats *Plecotus townsendii* and gray myotis *Myotis grisescens* flew more frequently through test frames at gated cave entrances with a round bar design or angle iron design than a funnel design (round bar: 40% of big-eared bats and 20% gray myotis exiting through the total cave entrance area, angle iron: 21% of big-eared bats and 16% of gray myotis, funnel: 7% of big-eared bats and 2% of gray myotis). Both caves had two entrances with existing gates. At one entrance at each site a 1 m² test frame was installed in front of the existing gate near the primary emergence pathway. Inserts of 3 different experimental designs were installed in the frames and tested for 5 consecutive nights each over 5 15 night periods (total of 25 nights per design). At dusk, bats were counted emerging through the frame and the remainder of the cave entrance. The total number of bats emerging during a 15-night period varied from 225 to 402 big-eared bats, and 1,065 to 2,649 gray myotis. The round bar design had 19 mm round steel bars in a 615 x 154 mm pattern. The angle iron design had 103 mm angle iron welded 154 mm apart in a horizontal pattern. The funnel was a 1 m² one-way metal funnel narrowing to an exit hole of 230 x 230 mm. Existing gates at the caves were of the round bar design.

A controlled, before-and-after study from 1994 to 1996 at a cave on a forested limestone ridge in north Florida, USA (4) found that significantly more southeastern myotis *Myotis austroriparius* and gray myotis *Myotis grisescens* emerged from a cave entrance when a steel bar gate was removed and replaced with a fence (average 306 bats/month, 8% of total bats emerging from cave with gate in place vs. average 1,517 bats/month, 48% after removal). The number of bats emerging from a second ungated entrance to the cave subsequently decreased (from 3,609 to 1,651 bats/month). Emerging bats were counted monthly at an open entrance and a gated entrance for one year before and one year after the cave gate was removed (August 1994 to July 1996). The cave gate consisted of steel bars 13 mm in diameter spaced 100 mm apart in one direction and 465 mm in the other. Before removal of the gate a 2.2 m high chain-link fence was erected 6–8 m from the cave entrance. The caves were originally gated in the 1970s after an increase in vandalism, but bat numbers were reported to decline after gating of the entrances.

A replicated, before-and-after study from 1981 to 2001 at six caves in a limestone plateau in north-eastern Oklahoma, USA (5) found that after the installation of cave gates the total number of gray myotis *Myotis grisescens* using the caves increased (from 60,130 bats in 1981 to 70,640 in 2001). Two caves had more gray myotis after gating, and three caves had no change in gray myotis numbers (one other cave had been gated prior to the study so pre- and post-gating estimates were not provided). Emergence times of the gray myotis colonies were determined in June and July 1999 and 2000 at three gated and three ungated caves but no difference was observed. Cave gates consisted of horizontal angle-iron bars with 150 mm spacing. Cave gates were gradually added to the entrances in different years throughout the study period. The sizes of the six gray myotis colonies were estimated during the summers 1981–1983, 1991, 1999 and 2001 from the size of guano accumulation.

In a replicated, controlled, site comparison study from July to October 2003 at 28 cave and mine sites from Ontario, Canada to Tennessee, USA (6) bats at cave entrances circled, retreated more and passed through less often with gates installed than at ungated entrances (37% of bats at existing gates, 60% at newly installed mock gates, and 23% at ungated entrances). Higher bat activity and smaller gate size (< 9.5 m²) increased this behaviour. Bat flight behaviour did not differ based on the vertical spacing of gate supports, the number of entrances or the position of the gate either at the cave entrance or an internal passageway. Echolocation calls and flight speed did not differ at gated entrances. Observations were made and echolocation calls and flight speeds recorded for multiple nights at 16 gated sites and 21 ungated sites. Nine of the gated sites were fitted with mock wooden gates (horizontal bars 25 mm diameter with 146 mm spacing) during the experiment. Existing cave gates were of a variety of designs. Between 30 and 257 bat flight behaviours were observed at each site. Captures and recording of echolocation calls revealed six to seven species across the sites.

In a replicated, controlled, before-and-after study in autumn 2003 at four derelict mines in a forested area near Eden in south-eastern Australia (7) the number of eastern horseshoe bats *Rhinolophus megaphyllus* and Schreiber's bats *Miniopterus schreibersii* at two caves were reduced after the installation of cave gates with a 125 mm horizontal spacing (from 540 to 290 bats and 120 to 30 bats, respectively). The number of bats aborting exit and entry flights also significantly increased. Horizontal spacing of 450 mm and 300 mm did not significantly affect bat numbers or behaviour. Bat numbers at two control caves either remained constant or increased. The four derelict mines were similar in internal height and complexity. Two of the mines were left ungated as controls, and two were fitted with template gates consisting of 20 mm plastic tubing. Activity at experimental mines was observed in stages of 11 days: pre-treatment followed by the successive addition of horizontal bars to reduce the spacing size (to 400, 300 and 125 mm). Bats were logged automatically using infrared beams, and observations of flight behaviour were made.

A before-and-after study from 1991 to 2004 at gated abandoned mines in forested areas of central and western Colorado, USA (8) found that 8 bat species continued to use 43 out of 47 mines with gates of various design up to 12 years after installation. Four types of gate were evaluated, all with bar spacings of 150 mm. Traditional gates allowed access to bats across the whole gate, ladder gates allowed access to bats at the centre only, and both types of gate were also constructed in metal culverts where mine entrances were too unstable to anchor the gate itself. None of the caves with full gates with or without culverts were abandoned by bats. Three caves with ladder gates and one cave with a culvert ladder gate were abandoned by bats. Single mines were surveyed 1–10 times with multiple methods (catching, visual counts and infrared motion detectors). The study looked at use of the caves by bats only, not at population sizes.

A controlled, before-and-after study from September to October in 2004 at a cave in a wooded limestone valley in northern England, UK (9) found that cave gates with horizontal bar spacings of 130 mm and 100 mm caused significantly more bats to abort their first and often subsequent attempts to enter the cave though the gate (proportion of bats entering/30 minute period with 130 mm spacing: 0.07 with gate, 0.2 before gate installation and 0.28 after gate removal; with 100 mm spacing: 0.08 with gate, 0.2 both before gate installation and after removal). Gates with horizontal spacings of 150 mm had no significant effect (proportion of bats entering/30 minute period: 0.1 with gate, 0.14 before installation and 0.16 after removal). Bat behaviour was found to return to normal after removal of the gates. The behaviour of swarming bats (predominantly Natterer's bats *Myotis nattereri*) was observed on 6–10 nights using night video recording, with gate size randomized between nights. One cave entrance was used for the experiments (1.5 m diameter) with custom-made gates (made with 15 mm diameter plastic tubing) of each of the 3 sizes positioned over it. Bats were recorded for three 30-minute periods with the gate open ('before'), closed, and open again ('after'). Long-term impacts of the gates were not investigated.

In a before-and-after study from 2002 to 2008 at the Dupnisa cave system in forested mountains in Turkey (10) the total number of 15 bat species within the system was found to significantly increase after it had been opened to controlled tourism with restrictions in place (maximum counts before 42,800 hibernating bats and 7,900 breeding/nursing bats vs. after 54,600 hibernating bats and 11,000 breeding/nursing bats). Two of the caves opened to tourists had an entrance gated using horizontal iron bars with 200 mm spacing. Other entrances inaccessible to humans were left ungated. The total number of bats in each of these caves increased after opening to tourism. A third cave, which remained closed to tourism, had constant numbers of bats throughout the study period. Population sizes were counted every 40 days for 1–2 days with 15 surveys before (2002–2004) and 38 surveys after opening to tourism (2004–2008). Before opening to tourism, recreational users had made frequent uncontrolled visits to the caves. After opening for tourism gates were installed on two entrances, daily and seasonal timings of visits were controlled by security guards, tourists were guided along set routes away from colonies with time limits for visits, information signs were erected, and lights were switched off outside of visiting times. The study does not distinguish between the effects of the cave gates and the different visiting restrictions imposed.

(1) Richter A.R., Humphrey S.R., Cope J.B. & Brack V. (1993) Modified cave entrances: thermal effect on body mass and resulting decline of endangered Indiana bats (*Myotis sodalis*). *Conservation Biology*, 7, 407–415.

(2) Voûte A.M. & Lina P.H.C. (1986) Management effects on bat hibernacula in the Netherlands. *Biological Conservation*, 38, 163–177.
(3) White D.H. & Seginak J.T. (1987) Cave gate designs for use in protecting endangered bats. *Wildlife Society Bulletin*, 15, 445–449.
(4) Ludlow M.E. & Gore J.A. (2000) Effects of a cave gate on emergence patterns of colonial bats. *Wildlife Society Bulletin*, 28, 191–196.
(5) Martin K.W., Leslie D.M., Payton M.E., Puckette W.L. & Hensley S.L. (2003) Internal cave gating for protection of colonies of the endangered gray bat (*Myotis grisescens*). *Acta Chiropterologica*, 5, 143–150.
(6) Spanjer G.R. & Fenton M.B. (2005) Behavioral responses of bats to gates at caves and mines. *Wildlife Society Bulletin*, 33, 1101–1112.
(7) Slade C. & Law B. (2008) An experimental test of gating derelict mines to conserve bat roost habitat in southeastern Australia. *Acta Chiropterologica*, 10, 367–376.
(8) Navo K.W. & Krabacher P. (2005) The use of bat gates at abandoned mines in Colorado. *Bat Research News*, 46, 1–8.
(9) Pugh M. & Altringham J.D. (2005) The effects of gates on cave entry by swarming bats. *Acta Chiropterologica*, 7, 293–300.
(10) Paksuz S. & Özkan B. (2012) The protection of the bat community in the Dupnisa Cave System, Turkey, following opening for tourism. *Oryx*, 46, 130–136.

7.2 Maintain microclimate at hibernation/roost sites

- We found no evidence for interventions that maintain the microclimate at roost sites.

See 'Use cave gates to restrict public access' for one study in which a stone wall and gate influenced the microclimate of a cave with an effect on hibernating bats.

7.3 Impose restrictions on cave visits

- Two before-and-after studies from Canada[1] and Turkey[2] found that bat populations within caves increased after restrictions on cave visitors were imposed.

Background
Cave visits by recreational users may be restricted to reduce disturbance to bat colonies. Examples of such restrictions are seasonal and daily timing of visits to avoid times when bats are vulnerable, closure to parts of caves close to bat colonies, time limits on visits, supervision of visitors by guides or security guards, and restrictions on the use of lights within caves. Often several restrictions will be used in conjunction and the individual effects of each cannot be distinguished.

A before-and-after study from 1983 to 2009 at Cadomin Cave in the Rocky Mountains, Canada (1) found significantly more hibernating bats after restrictions on visitors were enforced (average before approximately 450 bats/year vs. after 650 bats/year). Cadomin Cave was highly popular with recreational visitors. In 1997 seasonal access restrictions were imposed. In 1998 the area was established as a National Park and signs were erected to inform the public about cave access. Active enforcement to restrict recreational visitors in winter months began in 2000. An annual census of visual counts was carried out from 1983 to 2000, followed by a census every other year until 2009. Three bat species are known to use the cave, little brown myotis *Myotis lucifugus*, long-legged myotis *Myotis volans*, and low numbers of northern myotis *Myotis septentrionalis*.

In a before-and-after study from 2002 to 2008 at the Dupnisa cave system in forested mountains in Turkey (2), the total number of 15 bat species was found to significantly increase after the caves had been opened to controlled tourism with restrictions in place (maximum counts before 42,800 hibernating bats and 7,900 breeding/nursing bats vs. after 54,600 hibernating bats and 11,000 breeding/nursing bats). Two of the caves opened to tourists within the system had an entrance gated using horizontal iron bars with 200 mm spacing. Other entrances inaccessible to humans were left ungated. The total number of bats in each of these caves increased after opening to tourism. A third cave, which remained closed to tourism, had constant numbers of bats throughout the study period. Population sizes were counted every 40 days for 1–2 days with 15 surveys before (2002–2004) and 38 surveys after opening to tourism (2004–2008). Before opening to tourism, recreational users had made frequent uncontrolled visits to the caves. After opening for tourism gates were installed on two entrances, daily and seasonal timings of visits were controlled by security guards, tourists were guided along set routes away from colonies with time limits for visits, information signs were erected, and lights were switched off outside of visiting times. The study does not distinguish between the effects of the cave gates and the various visiting restrictions imposed.

(1) Olson C.R., Hobson D.P. & Pybus M.J. (2011) Changes in population size of bats at a hibernaculum in Alberta, Canada, in relation to cave disturbance and access restrictions. *Northwestern Naturalist*, 92, 224–230.

(2) Paksuz S. & Özkan B. (2012) The protection of the bat community in the Dupnisa Cave System, Turkey, following opening for tourism. *Oryx*, 46, 130–136.

7.4 Educate the public to reduce disturbance to hibernating bats

- We found no evidence for the effects of educating the public about bats to reduce disturbance at hibernation sites such as caves.

Background
Educational or informative signs for the public are often used at bat hibernation sites in conjunction with other interventions to reduce human disturbance. See 'Impose restrictions on cave visits'.

7.5 Legally protect bat hibernation sites

- We found no evidence for the effects of legally protecting bat hibernation sites that may be subject to human disturbance.

7.6 Provide artificial hibernacula for bats to replace disturbed sites

- We found no evidence for the effects of providing artificial hibernacula for bats to replace hibernation sites lost due to human disturbance.

8 Threat: Natural system modification – natural fire and fire suppression

Key messages

Use prescribed burning
Four studies in North America looked at bat activity and prescribed burning. One replicated, controlled, site comparison study found no difference in bat activity between burned and unburned forest. One replicated, site comparison study found higher activity of bat species that forage in the open in burned than unburned stands. One site comparison study found higher bat activity in forest preserves when prescribed burning was used with other restoration practices. One controlled, replicated, before-and-after study found that the home ranges of bats were closer to burned stands following fires. Four studies in North America (three replicated and one controlled) found bats roosting more often in burned areas, or equally in burned and unburned forest.

8.1 Use prescribed burning

- Four studies in North America looked at bat activity in forest stands with and without prescribed burning, with mixed results. One replicated, controlled, site comparison study[1] found no difference in bat activity between burned and unburned forest. One replicated, site comparison study[7] found higher activity of bat species that forage in the open in burned than unburned stands. One site comparison study[3] found higher bat activity in forest preserves when prescribed burning was used with other restoration practices. One controlled, replicated, before-and-after study[4] found that the home ranges of bats were closer to burned stands following fires.

- Four studies in North America (three replicated and one controlled) found bats roosting more often in burned areas[2,4], or equally in burned and unburned forest[5,6].

Background
Prescribed burning is a practice used in forest management where controlled burns are conducted to reduce the risk of more damaging uncontrolled natural fires and to stimulate tree germination. Controlled burning alters forest structure, opens up the tree canopy and creates potential roosts in snags.

Although there is evidence that prescribed or controlled burning can benefit bats, there may also be negative effects such as heat injury, smoke and carbon monoxide poisoning, and arousal from torpor. Consideration must be given to fire intensity, ignition procedures and seasonal timing of burnings (Dickinson *et al.* 2010).

Dickinson M.B., Norris J.C., Bova A.S., Kremens R.L., Young V. & Lacki M.J. (2010) Effects of wildland fire smoke on a tree-roosting bat: integrating a plume model, field measurements, and mammalian dose-response relationships. *Canadian Journal of Forest Research*, 40, 2187–2203.

A replicated, controlled, site comparison study in May–August 2001 and 2002 in an experimental pine forest in South Carolina, USA (1) found that in both years of the study overall bat activity was intermediate in burned and thinned/burned stands and not significantly different from control stands without any treatment (average 2.1 and 3.4 vs. 1.3 bat passes/night respectively in 2001, average 2 and 1.9 vs. 0.4 bat passes/night in 2002). Bat activity was the highest in thinned stands (average 8.2 bat passes/night in 2001, average 2.1 bat passes/night in 2002), but this difference was only statistically significant in 2001. Of the three most frequently recorded species, this pattern was consistent for big brown bats *Eptesicus fuscus* and eastern red bats *Lasiurus borealis*, but not for eastern pipistrelles *Perimyotis subflavus*, which did not vary between stand types in either year. Twelve 14 ha stands were selected with 3 replicates of 4 treatment types: prescribed burning (burned in April 2001 with strip head fire and flanking fires), thinning to 18 m^2/ha (in winter 2000–2001), thinning to 18 m^2/ha followed by prescribed burning (burned in spring 2002 with strip head fires) and a control with no treatment. Bat activity was sampled from sunset to sunrise with two bat detectors at random points in each stand for two nights per month from May–August in both years.

A replicated study in 2003–2004 in a 1,200 ha deciduous forest in Missouri, USA (2) found that evening bats *Nycticeius humeralis* roosted only in areas of the forest where prescribed burning had occurred. Twenty-three bats were tracked to 63 tree roosts in burned areas. This was significantly more than expected if burned or unburned forest were selected for roosting randomly. The burned area of the forest had higher light canopy penetration, lower canopy tree density and significantly more dead trees. Prescribed burning began in the study area in 1999 after 50 years of fire suppression. Burning was done every 2 years in March or April in 55% of the study area. Bats were caught from March 2003 to March 2004 using mist nets across forest roads between the burned and unburned areas of the forest and in ponds in both areas. Twenty-three bats (11 females and 12 males) were fitted with radio transmitters and tracked to roost trees each day until the transmitter was shed or expired. Vegetation sampling was done in random blocks or along transects in burned and unburned areas in 2003.

In a site comparison study in 2004–2005 in nine forest preserves within the Chicago metropolitan area, USA (3) the highest bat activity was recorded in two preserves that had undergone restoration with multiple prescribed burns, invasive plant species removal and snag recruitment (average 19 and 16 bat passes/preserve in 2004, average 7 and 18 bat passes/preserve in 2005). The lowest bat activity was recorded in a control site with no restoration (both years average one bat pass in total). Overall bat activity at all sites was positively related to prescribed burning, invasive species removal and small tree density (7.7–20 cm diameter at breast height) and negatively related to shrub density and clutter at heights of 0–6 m above the ground. The study does not distinguish between the effects of prescribed burning and the other restoration practices used. Responses to woodland restoration varied among bat species. The eastern red bat *Lasiurus borealis* was positively associated with small and medium (20–33 cm) tree densities and negatively related to clutter at 0–9 m. *Myotis* spp. were positively related with canopy cover, clutter at 6–9 m and small and medium tree densities. The silver-haired bat *Lasionycteris noctivagans* was positively associated with more open forests. Activity of the big brown bat *Eptesicus fuscus* was not associated with any vegetation variables in the study. The nine forest preserves varied in size from 10 to 260 ha. Fire suppression over the last 100 years had altered the structural diversity of the forests. Eight of the forest preserves were under management to restore forest to pre-European settlement conditions. Restoration practices were used to open canopy cover, reduce tree density and remove invasive plant species. Bats were monitored for four hours from sunset with bat detectors in June–September 2004 and May–August 2005 for five nights per site per year. Twenty randomly located 30 m line transects were sampled per site with 4 detectors placed 10 m apart along each transect.

In a replicated, controlled, before-and-after study in 2006 and 2007 in three sites in a mixed forest in a river gorge in Kentucky, USA (4) more female northern myotis *Myotis septentrionalis* roosted in burned habitats than unburned habitats following prescribed fires, although the difference in the number of roosts in each habitat type was not statistically tested (74%, 26 roosts in burned habitat vs. 26%, 9 roosts in unburned). In both habitats, bats selected tall roost trees in the early stages of decay. After burning, bats chose roosts in trees with a greater number of cavities and more exfoliating bark. The size of home ranges (average 60 ha before burning and 72 ha after) and core areas (average 11 ha before burning and 14 ha after) did not vary significantly between bats radio-tracked before and after fires, but home ranges were closer to burned habitats than unburned habitats following fires. The abundance of insects increased significantly after prescribed fire (average from 140 to 188 insects/trap/night). Two sites (435 ha and 185 ha) that were previously unburned were subject to prescribed burning in April 2007, with 54% of the area burned (mainly ridges and upper slopes). The third site (2,400 ha) was left unburned. Bats were captured from June–July 2006 and April–September 2007 using mist nets over ponds in burned and unburned sites. Radio transmitters were attached to 18 adult female bats, which were tracked nightly while foraging and to roost trees until the transmitter expired or was shed (average of 6 days). Characteristics of roost trees were compared to random snags. Insects were sampled with light traps from dusk until dawn at four locations in each burned site before and after burning.

A replicated, controlled, site comparison study in 2007 and 2008 in a 1,900 ha experimental forest in West Virginia, USA (5) found female northern myotis *Myotis septentrionalis* roosting in stands treated with prescribed fire and in unburned control stands (25 and 44 roosts respectively). The difference in the number of roosts in each stand type was not statistically tested. In both stand types, the majority of roosts (60% in burned and 62% in unburned stands) were in decaying trees with loose bark. Roost trees in burned stands were associated with larger canopy gaps than roost trees in control stands. A higher proportion of roost trees were found to be available in the burned stand. There was no significant difference in the frequency of roost switching (1–6 days in burned and 1–5 days in unburned stands) or the distance between roost trees (average 152 m in burned and 230 m in unburned stands). In April or May 2007 and 2008, 3 stands (45, 13 and 21 ha) were subjected to prescribed burning for one day using a strip head fire technique. Control stands were considered to be any area in the forest outside of the burned stands. In May–August 2007 and 2008 bats were captured over streams, pools, trails and service roads at burned and control sites using mist nets, and radio-transmitters were fitted. In 2007, three adult female bats were tracked to eight roosts. In 2008, 33 bats were tracked to 65 roosts, 4 of which were used previously in 2007.

In a replicated, controlled, site comparison study in 2007 and 2008 in two forests in West Virginia, USA (6), 15 male Indiana bats *Myotis sodalis* were tracked to 16 roosts in burned areas and 34 roosts in unburned areas. The difference in the number of roosts in each stand type was not statistically tested. In burned areas bats roosted in snags killed by fire, and in unburned areas in live trees. Roost trees in burned stands were surrounded by larger canopy gaps than random roost trees or roosts in unburned areas. There was no significant difference in the frequency of roost switching (1–4 days in burned and 1–2 days in unburned stands) or the distance between roost trees (average 220 m in burned and 477 m in unburned stands). In April or May 2007–2009 three stands (12, 13 and 121 ha) within one of the forests (Fernow Experimental Forest, 1,900 ha) were subjected to prescribed burning using a strip head fire technique. In the other forest site (Petit Farm, 400 ha) in March 2003 an escaped campfire had burned part of the forest stand. Control stands were unburned areas in each forest. Bats were captured over streams, pools, ponds and trails using mist nets and fitted with radio transmitters in summer 2004–2006 at Petit Farm and summer 2008–2009 at Fernow Experimental Forest, and also at a cave swarming site at Fernow Experimental Forest in autumn 2007–2008. Habitat variables were recorded using a point-quarter sampling method, and at random trees in burned and unburned areas.

In a replicated, site comparison study in 2008–2009 in two fire managed mixed forests in Florida, USA (7), bat species with high wing loading and aspect ratios (less manoeuvrable species that forage in open habitats, e.g. big brown bats *Eptesicus fuscus*) had significantly higher activity in the understorey of forest sites with a higher burn frequency (average 0.9 bat passes/site/night for 1–2 year burn, 0.5 bat passes/site/night for 3–5 year burn, 0.1 bat passes/site/night for > 8 year burn). The activity of eastern red bats *Lasiurus borealis* and Seminole bats *Lasiurus seminolus* (which could not be classified according to wing shape) followed the same pattern. The activity of bat species with low wing loading and aspect ratios (more manoeuvrable species that forage in cluttered habitats, e.g. southeastern bat *Myotis austroriparius*) did not differ significantly between burn treatments. Activity above the canopy did not differ between burn treatments for any species. Increased fire frequency resulted in a significant reduction in basal area, canopy density, snag density, shrub density and leaf litter, and a significant increase in canopy closure height, and grass and sand-ash ground cover. The abundance of all insects significantly increased with burn frequency, except for Lepidoptera which decreased. Twenty-four 40 ha square study plots were randomly selected in each forest with different burn frequencies: burned within the previous year with a burn frequency of 1–2 years, last burned 3–5 years prior and with a burn frequency of 3–5 years, burned more than 8 years prior with a burn frequency of > 8 years. Bat activity was recorded with remote detectors below the canopy (and above the canopy for 1–2 year and > 8 year burn frequencies) from 30 minutes before sunset to 30 minutes after sunrise from May–August in both years. Detectors were placed at two randomly chosen sites per burn category per night for four evenings per week. Vegetation surveys were conducted in both years in September and October along transects at four 15 m radius circular plots at each study site. Insects were sampled with light traps from dusk until dawn at two or three sites per week from May–August in both years.

(1) Loeb S.C. & Waldrop T.A. (2008) Bat activity in relation to fire and fire surrogate treatments in southern pine stands. *Forest Ecology and Management*, 255, 3185–3192.
(2) Boyles J.G. & Aubrey D.P. (2006) Managing forests with prescribed fire: implications for a cavity-dwelling bat species. *Forest Ecology and Management*, 222, 108–115.
(3) Smith D.A. & Gehrt S.D. (2010) Bat response to woodland restoration within urban forest fragments. *Restoration Ecology*, 18, 914–923.
(4) Lacki M.J., Cox D.R., Dodd L.E. & Dickinson M.B. (2009) Response of northern bats (*Myotis septentrionalis*) to prescribed fires in eastern Kentucky forests. *Journal of Mammalogy*, 90, 1165–1175.
(5) Johnson J.B., Edwards J.W., Ford W.M. & Gates J.E. (2009) Roost tree selection by northern myotis (*Myotis septentrionalis*) maternity colonies following prescribed fire in a Central Appalachian Mountains hardwood forest. *Forest Ecology and Management*, 258, 233–242.
(6) Johnson J.B., Ford W.M., Rodrigue J.L., Edwards J.W. & Johnson C.M. (2010) Roost selection by male Indiana myotis following forest fires in Central Appalachian hardwoods forests. *Journal of Fish and Wildlife Management*, 1, 111–121.
(7) Armitage D.W. & Ober H.K. (2012) The effects of prescribed fire on bat communities in the longleaf pine sandhills ecosystem. *Journal of Mammalogy*, 93, 102–114.

9 Threat: Invasive species and disease

Key messages – Invasive species

Remove invasive plant species
One site comparison study in North America found higher bat activity in forest preserves where invasive plant species had been removed alongside other restoration practices.

Control invasive predators
We found no evidence for the effects of controlling invasive species that are predators to bats.

Translocate to predator or disease free areas
Two small unreplicated studies in New Zealand and Switzerland found low numbers of bats remaining at release sites after translocation, and observed homing tendencies, disease and death.

Key messages – White-nose syndrome

Control anthropogenic spread
We found no evidence for the effects of interventions to control the anthropogenic spread of white-nose syndrome to new areas.

Increase population resistance
We found no evidence for the effects of increasing the resistance to white-nose syndrome of wild bat populations.

Cull infected bats
We found no evidence for the effects of culling bats infected with white-nose syndrome.

Modify cave environments to increase bat survival
We found no evidence for the effects of modifying cave environments to reduce the spread of white-nose syndrome and improve the survival of infected bats.

Invasive species

9.1 Remove invasive plant species

- One site comparison study[1] in North America found higher bat activity in forest preserves where invasive plant species had been removed alongside other restoration practices.

Background
Invasive plant species can threaten native biodiversity and alter bat foraging habitats such as forest and woodland. For example, invasive tree and vine species have caused

the deterioration of foraging habitat of the Seychelles sheath-tailed bat *Coleura seychellensis* and have been found to obstruct roost entrances (Gerlach 2009).

Gerlach, J. (2009) Conservation of the Seychelles sheath-tailed bat *Coleura seychellensis* on Silhouette Island, Seychelles. *Endangered Species Research*, 8, 5–13.

In a site comparison study in 2004–2005 in 9 forest preserves within the Chicago metropolitan area, USA (1) the highest bat activity was recorded in 2 preserves that had undergone restoration with multiple prescribed burns, invasive plant species removal and snag recruitment (average 19 and 16 bat passes/preserve in 2004, average 7 and 18 bat passes/preserve in 2005). The lowest bat activity was recorded in a control site with no restoration (both years average one bat pass in total). Overall bat activity at all sites was positively related to prescribed burning, invasive species removal and small tree density (7.7–20 cm diameter at breast height) and negatively related to shrub density and clutter at heights of 0–6 m above the ground. The study does not distinguish between the effects of removing invasive plant species and the other restoration practices used. Responses to woodland restoration varied among bat species. The eastern red bat *Lasiurus borealis* was positively associated with small and medium (20–33 cm) tree densities and negatively related to clutter at 0–9 m. *Myotis* spp. were positively related with canopy cover, clutter at 6–9 m and small and medium tree densities. The silver-haired bat *Lasionycteris notivagans* was positively associated with more open forests. Activity of the big brown bat *Eptesicus fuscus* was not associated with any vegetation variables in the study. The 9 forest preserves varied in size from 10 to 260 ha. Fire suppression over the last 100 years had altered the structural diversity of the forests. Eight of the forest preserves were under management to restore forest to pre-European settlement conditions. Restoration practices were used to open canopy cover, reduce tree density and remove invasive plant species. Bats were monitored for four hours from sunset with bat detectors in June–September 2004 and May–August 2005 for five nights per site per year. Twenty randomly located 30 m line transects were sampled per site with 4 detectors placed 10 m apart along each transect.

(1) Smith D.A. & Gehrt S.D. (2010) Bat response to woodland restoration within urban forest fragments. *Restoration Ecology*, 18, 914–923.

9.2 Control invasive predators

- We found no evidence for the effects of controlling invasive species that are predators to bats.

Background
Introduced predators such as rats, feral cats and snakes can threaten bat populations. The brown tree snake *Boiga irregularis* which invaded Guam in the 1950s was responsible for the extermination of two bat species. Eradication programmes have been successful, for example with rats *Rattus* spp. on the San Jorge Islands, Mexico (Donlan et al. 2003), but we found no evidence for the subsequent effects on local bat populations.

Donlan C.J., Howald G.R., Tershy B.R. & Croll D.A. (2003) Evaluating alternative rodenticides for island conservation: roof rat eradication from the San Jorge Islands, Mexico. *Biological Conservation*, 29–34.

9.3 Translocate to predator or disease free areas

- Two small, unreplicated studies in New Zealand[1] and Switzerland[2] found low numbers of bats remaining at release sites after translocation, and observed homing tendencies, disease and death.

Background
The translocation of bats involves the transport and release of bats from one area to another. This may be done to protect bats against threats from introduced predators, competitors or disease. Previous studies on the homing behaviour of bats have shown that bats will often attempt to fly long distances to return home when released in new areas (e.g. Davis & Cockrum 1962).

Davis R. & Cockrum E.L. (1962) Repeated homing exhibited by a female pallid bat. *Science*, 137, 341–342.

In a single study in 2005 on Kapiti Island, New Zealand (1) 9 out of 20 translocated lesser short-tailed bats *Mystacina tuberculata* were recorded at the release site 232 days after release. After eight months, captured bats were balding and had damaged, infected ears and were subsequently returned to captivity. Four male and 16 female captive bred juveniles were released in April and provided with roosts and supplementary food (fed consistently for 55 days after release and irregularly for 156 days after release). Bats were monitored using infrared video cameras, and caught in harp traps during three study periods after release (eight weeks in April–June, five weeks in August–September, one week in November–December). Kapiti Island is a 1,965 ha nature reserve of forest and scrub located 40 km south-west from the source bat population on mainland New Zealand.

In a small translocation study in 2006 in alpine villages surrounded by mountains, woodland and farmland in Switzerland (2) 10 out of 13 greater horseshoe bats *Rhinolophus ferrumequinum* and lesser horseshoe bats *Rhinolophus hipposideros* released more than 20 km from original roosts homed. Five greater and lesser horseshoe bats released more than 40 km from original roosts did not show homing tendencies. Within three days of release one greater horseshoe bat and three lesser horseshoe bats died. Two of the greater horseshoe bats translocated long distances settled in the release area and one female was regularly observed in its new roost in 2007 and 2008. Male and female bats of three age classes (adult, one to two years and yearlings) were captured from large colonies and translocated to small relict colonies in similar habitats between 18 and 148 km away. Released bats were monitored with infrared video and radio tracking.

(1) Ruffell J. & Parsons S. (2009) Assessment of the short-term success of a translocation of lesser short-tailed bats (*Mystacina tuberculata*). *Endangered Species Research*, 8, 33–39.
(2) Weinberger I.C., Bontadina F. & Arlettaz R. (2009) Translocation as a conservation tool to supplement relict bat colonies: a pioneer study with endangered horseshoe bats. *Endangered Species Research*, 8, 41–48.

White-nose syndrome

Background
White-nose syndrome is a condition in which a fungus *Pseudogymnoascus destructans* invades the skin around the muzzle and wings of hibernating bats. Infection causes bats to rouse from torpor more frequently and for longer periods, using up vital fat reserves and resulting in death. The disease has spread rapidly across North America

and is responsible for the deaths of millions of bats. No obvious treatment or means of preventing transmission is currently known, and most of the interventions below are in action as precautionary measures.

9.4 Control anthropogenic spread

- We found no evidence for the effects of interventions to control the anthropogenic spread of white-nose syndrome to new areas.

Background
This intervention involves restricting human access to caves and the decontamination of clothing and equipment.

9.5 Increase population resistance

- We found no evidence for the effects of increasing the resistance to white-nose syndrome of wild bat populations.

Background
This intervention involves treating or immunizing a proportion of bats within the population for white-nose syndrome to reduce the level and spread of the disease. There are practical issues for the delivery of treatments, and an effective vaccination has yet to be developed.

9.6 Cull infected bats

- We found no evidence for the effects of culling bats infected with white-nose syndrome.

Background
Culling of bats infected with white-nose syndrome has been considered to reduce transmission and slow the spread of the disease. However, this has not been tested and simulation modelling indicates that culling will not be an effective method to control the spread of white-nose syndrome (Hallam & McCracken 2011).

Hallam T.G. & McCracken G.F. (2011) Management of the panzootic white-nose syndrome through culling of bats. *Conservation Biology*, 25, 189–194.

9.7 Modify cave environments to increase bat survival

- We found no evidence for the effects of modifying cave environments to reduce the spread of white-nose syndrome and improve the survival of infected bats.

9.7 Modify cave environments to increase bat survival

Background
The fungus responsible for white-nose syndrome *Pseudogymnoascus destructans* grows in cold temperatures, and artificial heating of affected caves may slow fungus growth and improve the survival of infected bats by reducing the energy required for them to rouse from torpor. The growth and performance of *P. destructans* (previously called *Geomyces destructans*) in laboratory conditions has been found to decline rapidly over 15°C (Verant *et al.* 2012).

Verant M.L., Boyles J.G., Waldrep W. Jr., Wibbelt G. & Blehert D.S. (2012) Temperature-dependent growth of *Geomyces destructans*, the fungus that causes bat white-nose syndrome. *PLoS ONE*, 7, e46280.

10 Threat: Pollution

Key messages – Domestic and urban waste water

Change effluent treatments
We found no evidence for the effects on bats of changing effluent treatments of waste water discharged into rivers. One replicated, site comparison study in the UK found that foraging activity over filter bed sewage treatment works was higher than activity over active sludge systems.

Key messages – Agricultural and forestry effluents

Introduce legislation to control use
We found no evidence for the effects of introducing legislation to control the use of chemicals such as fertilizers, insecticides and pesticides.

Change effluent treatments
We found no evidence for the effects of changing the effluent treatments used in agriculture or forestry.

For evidence relating to reducing chemical use on farms see 'Threat: Agriculture – Intensive farming – Convert to organic farming'.

Key messages – Light and noise pollution

Leave bat roosts, roost entrances and commuting routes unlit
Two replicated studies in the UK found more bats emerging from roosts or flying along hedgerows when left unlit than when illuminated with white lights or streetlamps.

Minimize excess light pollution
One replicated, randomized, controlled study in the UK found that bats avoided flying along hedgerows with dimmed lighting, and activity levels were lower than along unlit hedges. We found no evidence for the effects on bats of reducing light spill using directional lighting or hoods.

Restrict timing of lighting
We found no evidence for the effects of limiting the operational hours of lighting to reduce disturbance to bats.

Use low pressure sodium lamps or use UV filters
We found no evidence for the effects on bats of using low pressure sodium lamps or lights with UV filters.

Impose noise limits in proximity to roosts and bat habitats

We found no evidence for the effects of imposing noise limits in proximity to bat roosts or important bat habitats.

Key messages – Timber treatments

Use mammal safe timber treatments in roof spaces

Two controlled laboratory studies in the UK found commercial timber treatments (containing lindane and pentachlorophenol) to be lethal to bats, but found alternative artificial insecticides (including permethrin) and three other fungicides did not increase bat mortality. Sealants over timber treatments had varying success.

Restrict timing of treatment

One controlled laboratory experiment in the UK found that treating timber with lindane and pentachlorophenol 14 months prior to exposure by bats increased survival time but did not prevent death. Bats in cages treated with permethrin survived just as long when treatments were applied 2 months or 14 months prior to exposure.

Domestic and urban waste water

10.1 Change effluent treatments

- We found no evidence for the effects on bats of changing effluent treatments of waste water discharged into rivers. One replicated, site comparison study in the UK[1] found that foraging activity over filter bed sewage treatment works was higher than activity over active sludge systems.

Background
Organic pollution occurs when treated sewage effluents containing organic compounds are discharged into rivers affecting plant growth and the number and diversity of insects. Riparian habitats are important for foraging bats and changes in water quality may have positive effects for some species, and negative effects for others (Vaughan *et al.* 1996, Kalcounis-Rüppell *et al.* 2007, Abbott *et al.* 2009).
We found evidence that filter sewage bed treatment works can provide foraging habitat for bats. However, the results must be treated with caution as a subsequent study found that insects above these filter beds were contaminated with endocrine-disrupting chemicals that may have adverse effects on bats feeding on them (Park *et al.* 2009).

Abbott I.M., Sleeman D.P. & Harrison S. (2009) Bat activity affected by sewage effluent in Irish rivers. *Biological Conservation*, 142, 2904–2914.
Kalcounis-Rüppell M.C., Payne V., Huff S.R. & Boyko A. (2007) Effects of wastewater treatment plant effluent on bat foraging ecology in an urban stream system. *Biological Conservation*, 138, 120–130.
Park K.J., Müller C.T., Markman S., Swinscow-Hall O., Pascoe D. & Buchanan K.L. (2009) Detection of endocrine disrupting chemicals in aerial invertebrates at sewage treatment works. *Chemosphere*, 77, 1459–1464.
Vaughan N., Jones G. & Harris S. (1996) Effects of sewage effluent on the activity of bats (Chiroptera: Vespertilionidae) foraging along rivers. *Biological Conservation*, 78, 337–343.

In a replicated, site comparison study between June and August 2003 at 30 sewage treatment works in central and southern Scotland, UK (1) significantly higher activity of *Pipistrellus* spp.

was recorded over percolating filter beds than over activated sludge systems (average 15 vs. 4 bat passes/15 min respectively). Foraging activity of *Pipistrellus* spp. over filter beds was comparable to that at nearby foraging habitat along river banks, whereas foraging activity over activated sludge sites was lower. At filter beds, waste water is sprayed over inert filter material creating a microbial film which supports high insect numbers. In activated sludge systems, sewage and bacterial sludge are mixed together creating an unfavourable habitat for insects. Insect biomass was found to be significantly higher at filter beds. Bat activity was recorded with bat detectors at 3 points per site for 15 minutes each after dusk, and insects were assessed using suction traps for 2 hours after dusk at each site. *Myotis* spp. were also detected at both types of treatment works but numbers were too low for analysis.

(1) Park K.J. & Cristinacce A. (2006) Use of sewage treatment works as foraging sites by insectivorous bats. *Animal Conservation*, 9, 259–268.

Agricultural and forestry effluents

10.2 Introduce legislation to control use

- We found no evidence for the effects of introducing legislation to control the use of chemicals such as fertilizers, insecticides and pesticides.

10.3 Change effluent treatments

- We found no evidence for the effects of changing the effluent treatments used in agriculture or forestry.

Light and noise pollution

Background
Light and noise pollution may disturb bats and degrade foraging habitat. Some bat species avoid lit areas (e.g. Stone et al. 2009, 2012), whereas others are attracted to street lights to forage (e.g. Rydell 1992, Blake et al. 1994) putting them at risk of predation or collisions with traffic. Traffic noise has been found to reduce the foraging success of bats that use passive listening to hunt (Siemers & Schaub 2010), and noise and lighting at a music festival resulted in bats emerging later from roosts (Shirley et al. 2001).

Blake D., Hutson A.M., Racey P.A., Rydell J. & Speakman J.R. (1994) Use of lamplit roads by foraging bats in southern England. *Journal of Zoology*, 234, 453–462.
Rydell J. (1992) Exploitation of insects around streetlamps by bats in Sweden. *Functional Ecology*, 6, 744–750.
Shirley M.D.F., Armitage V.L., Barden T.L., Gough M., Lurz P.W.W., Oatway D.E., South A.B. & Rushton S.P. (2001) Assessing the impact of a music festival on the emergence behaviour of a breeding colony of Daubenton's bats (*Myotis daubentonii*). *Journal of Zoology*, 254, 367–373.
Siemers B.M. & Schaub A. (2010) Hunting at the highway: traffic noise reduces foraging efficiency in acoustic predators. *Proceedings of the Royal Society B-Biological Sciences*.
Stone E.L., Jones G. & Harris S. (2009) Street lighting disturbs commuting bats. *Current Biology*, 19, 1–5.
Stone E.L., Jones G. & Harris S. (2012) Conserving energy at a cost to biodiversity? Impacts of LED lighting on bats. *Global Change Biology*, 18, 2458–2465.

10.4 Leave bat roosts, roost entrances and commuting routes unlit

- Two replicated studies in the UK found more bats emerging from roosts[1] or flying along hedgerows[2] when left unlit than when illuminated with white lights or streetlamps.

In a small, replicated, controlled study in July and August 2000 at two bat roosts in buildings near woodland in Aberdeenshire, UK (1) significantly fewer soprano pipistrelles *Pipistrellus pygmaeus* emerged when roosts were lit with white lights than when roosts were unlit (average 2 vs. 40 bats at roost 1 respectively, and 24 vs. 90 bats at roost 2). Fewer bats also emerged with blue lights at both roosts (average 6 bats at roost 1 and 62 bats at roost 2) and with the red light at roost 1 (average 13 bats). There was no significant difference between the number of bats emerging with the red light (average 72 bats) and the unlit treatment at roost 2. A hand-held halogen light with coloured filters was placed within 3 m of roost 1 and 5 m of roost 2. The number of bats emerging per 30 second interval was counted at dusk from the 2 roosts over 20 nights. Each unlit control night was followed by an experimental night with white, red and blue lights rotated in a random order and changed every 30 seconds.

In a replicated, controlled study from April to July 2008 along eight hedgerows in the south of the UK (2) the activity of lesser horseshoe bats *Rhinolophus hipposideros* was significantly lower along lit than unlit hedges (average 10 vs. 100 bat passes respectively). Lesser horseshoe bats became active significantly later on nights when hedges were lit (average 89 min after sunset) than nights when they were unlit (average 30 min after sunset). Hedges were illuminated with 2 portable high pressure sodium streetlights with average light levels of 53 lux. Bat activity was recorded with bat detectors and observations of behaviour were made. Experiments were conducted for seven nights per site with a silent unlit control treatment for one night, a noise treatment on the second night (with the generator powering the lights), four nights with the lit treatment and a final night with a repeat of the noise treatment. Generator noise did not affect bat activity levels, and did not delay the initiation of activity.

(1) Downs N.C., Beaton V., Guest J., Polanski J., Robinson S.L. & Racey P.A. (2003) The effects of illuminating the roost entrance on the emergence behaviour of *Pipistrellus pygmaeus*. *Biological Conservation*, 111, 247–252.
(2) Stone E.L., Jones G. & Harris S. (2012) Conserving energy at a cost to biodiversity? Impacts of LED lighting on bats. *Global Change Biology*, 18, 2458–2465.

10.5 Minimize light pollution

- One replicated, randomized, controlled study in the UK[1] found that bats avoided flying along hedgerows with dimmed lighting, and activity levels were lower than along unlit hedges. We found no evidence for the effects on bats of reducing light spill using directional lighting or hoods.

> **Background**
> Light pollution may be minimized by reducing light levels (e.g. low wattage or low intensity lights) and reducing light spill (by using directional lighting or hoods).

In a replicated, randomized, controlled study along ten hedgerows in south-west England and Wales, UK (1) the activity of lesser horseshoe bats *Rhinolophus hipposideros* and *Myotis* spp. was lower when hedges were lit with LED lights of three different intensities compared with unlit control trials. For *Myotis* spp. there was no difference in activity between high, medium and low light treatments (average 5 bat passes for each light treatment, 35 bat passes when unlit). For lesser horseshoe bats, activity was significantly lower when hedges were lit with high intensity lights than with medium or low light treatments (average 5 bat passes for high light, 22 for medium light, and 37 for low light), but even low intensity lights caused a significant decline in activity compared to unlit control trials (average 100 bat passes). Both species were observed avoiding the lights. Hedges were illuminated with LED street lights (consisting of 24 x 2.4 watt high power LEDs). Experiments were conducted for six nights per site with five different treatments: a silent unlit control treatment, a noise treatment repeated twice (with the generator powering the lights) and three lit treatments in a randomized order of low (3.6 lux), medium (6.6 lux) and high intensity (49.8 lux). Bat activity was recorded with bat detectors and observations of behaviour were made. *Pipistrellus*, *Nyctalus* and *Eptesicus* spp. were also observed but were not significantly affected by the lighting.

(1) Stone E.L., Jones G. & Harris S. (2012) Conserving energy at a cost to biodiversity? Impacts of LED lighting on bats. *Global Change Biology*, 18, 2458–2465.

10.6 Restrict timing of lighting

- We found no evidence for the effects of limiting the operational hours of lighting to reduce disturbance to bats.

10.7 Use low pressure sodium lamps or use UV filters

- We found no evidence for the effects on bats of using low pressure sodium lamps or UV filters on lights.

Background
Insects are attracted to lights that emit an ultraviolet (UV) component. This may draw insects away from bat foraging habitats and attract some bat species to forage under streetlights. Low pressure sodium lamps emit light with low UV levels. UV filters may be used to filter out the UV component of lights.

10.8 Impose noise limits in proximity to roosts and bat habitats

- We found no evidence for the effects of imposing noise limits in proximity to bat roosts or important bat habitats.

Timber treatments

Background
Chemicals such as insecticides and fungicides are often applied to roof timbers in buildings where bats roost, to protect against wood-boring beetles and wood-rotting fungus. The increased use of chemicals, usually chlorinated hydrocarbons such as lindane (or gamma HCH) and pentachlorophenol (PCP) were linked with declines in bat populations in the 1980s (Stebbings & Griffith 1986).

Lindane has been found to be lethal to bats (Boyd *et al*. 1988) and is now rarely used. There is a statutory requirement that timber treatments containing lindane are labelled as 'dangerous to bats'. In the UK, 'mammal-safe' timber treatments are now widely available and are regulated by the Health and Safety Executive with strict directions for use.

Most early studies of the toxicity of timber treatments were on laboratory rodents. Only a few studies exist which provide evidence for the effect of different timber treatments on bats.

Boyd I.L., Myhill D.G. & Mitchell-Jones A.J. (1988) Uptake of gamma-HCH (lindane) by pipistrelle bats and its effect on survival. *Environmental Pollution*, 51, 95–111.

Stebbings R.E. & Griffith F. (1986) *Distribution and Status of Bats in Europe*. Institute of Terrestrial Ecology, Huntingdon, UK.

10.9 Use mammal-safe timber treatments in roof spaces

- Two controlled laboratory studies (one replicated) in the UK[1, 2] found a commercial remedial timber treatment containing insecticides (lindane) and fungicides (pentachlorophenol) to be lethal to bats in conditions that simulated exposure in the wild. Artificial pyrethroid insecticides (including permethrin) and three other fungicides did not increase bat mortality. One of the studies found that one of two sealants used over timber treatments prevented death[1].

In a controlled laboratory study in the summers of 1982–1984 in south-west Scotland, UK (1) all common pipistrelle bats *Pipistrellus pipistrellus* placed in experimental cages treated with a commercial remedial timber treatment died (containing 1% w/v lindane and 5% w/v pentachlorophenol in an organic solvent). A layer of polyurethane varnish over chlorinated hydrocarbon timber treatments did not prevent death, but an application of acrylic resin did. Bats in cages treated with artificial pyrethroid insecticides (permethrin 0.3%, cypermethrin 0.05% or deltamethrin 0.02%) did not differ in survival from bats in untreated control cages and all went on to reproduce successfully. Bats in cages treated with the alternative fungicide tributyltin oxide had high mortality. The other fungicides tested (copper naphthenate 2.76%, borester 5–15% and zinc octoate 8%) did not result in higher mortality than control groups. All bats used in the experiment were female common pipistrelles caught at nursery roosts and fed with mealworms for three weeks prior to the experiments. Cages were 40 x 20 x 20 cm made from steel or zinc and lined with plywood. Treated cages had chemicals applied at a rate of 0.5 litre m^{-2}. Cages were kept in unheated rooms with constant conditions, and bats were provided with food, water and vitamins. Conditions simulated the level of exposure to the chemicals that bats would experience in the wild. The duration of experiments varied from 72–154 days, and 8–14 bats were used in each trial.

In a replicated, controlled, laboratory experiment from July to October 1988 in Suffolk, UK (2) pipistrelle bats *Pipistrellus* spp. roosting in boxes treated with permethrin (0.2% w/w solution) survived as well as bats in control boxes treated with a solvent only (7% w/w ethanol in white spirit). Nine out of ten bats survived in each treatment after 32 days. All bats in boxes treated with pentachlorophenol (PCP, 5% w/w solution) or a mixture of PCP (5% w/w) and permethrin (0.2% w/w) died within 24 hours and 120 hours respectively. Surface concentrations from wood scrapings were 65 mg g^{-1} for PCP only, 74 mg g^{-1} for the PCP/permethrin mixture and 3 mg g^{-1} for permethrin only. Pipistrelle bats were caught from a nursery roost in April and kept in captivity on a diet of mealworms, vitamins and water. In October, 4 groups of 9 to 10 bats were put in outdoor flight enclosures with roost boxes (12 x 4 x 12 cm) made from 24 mm timber treated with each of the four treatment types. Pesticides were formulated in white spirit as in commercially available timber treatments. Treated boxes were dried for a total of 96 hours before use. Bats were checked daily, and the experiments were terminated after 32 days. Post-mortems were carried out and PCP was detected in the body tissues of dead bats.

(1) Racey P.A. & Swift S.M. (1986) Residual effects of remedial timber treatments on bats. *Biological Conservation*, 35, 205–214.
(2) Shore R.F., Myhill D.G., French M.C., Leach D.V. & Stebbings R.E. (1991) Toxicity and tissue distribution of pentachlorophenol and permethrin in pipistrelle bats experimentally exposed to treated timber. *Environmental Pollution*, 73, 101–118.

10.10 Restrict timing of timber treatment application

- One controlled laboratory experiment in the UK[1] found that treating timber with lindane and pentachlorophenol 14 months prior to exposure by bats increased survival but did not prevent death. Bats in cages treated with permethrin survived just as long when treatments were applied 2 months or 14 months prior to exposure.

In a controlled laboratory study during the summers of 1982–1984 in south-west Scotland, UK (1) all common pipistrelle bats *Pipistrellus pipistrellus* placed in experimental cages treated with a commercial remedial timber treatment died (containing 1% w/v lindane and 5% w/v pentachlorophenol in an organic solvent). Bats survived longer in cages that had been treated 14 months previously (average 15 days) than cages treated 6 weeks previously (average 4 days), but all bats still died. Bats in cages treated with permethrin (0.3%) did not differ in survival from bats in untreated control cages and the timing of application had no effect on survival (seven or eight out of ten bats survived the duration of the experiments). All bats used in the experiment were female common pipistrelles caught at nursery roosts and fed with mealworms for three weeks prior to the experiments. Cages were 40 x 20 x 20 cm made from steel or zinc and lined with plywood. One cage was used per treatment and chemicals were applied at a rate of 0.5 litre m^{-2}. Cages were kept in unheated rooms with constant conditions, and bats were provided with food, water and vitamins. Conditions simulated the level of exposure to the chemicals that bats would experience in the wild. The duration of experiments was 113 or 120 days, and 10 or 14 bats were used in each trial.

(1) Racey P.A. & Swift S.M. (1986) Residual effects of remedial timber treatments on bats. *Biological Conservation*, 35, 205–214.

11 Providing artificial roost structures for bats

Background

Bats roost in caves, built structures, natural crevices (e.g. in rocks) and in trees. The provision of artificial roost structures for bats is a widely used intervention, as a conservation measure and for research, and there is a lot of literature on the use of these structures by bats. However, the many different designs of artificial roost structures available makes it difficult to draw consistent conclusions as evidence in support of each individual design is lacking. We particularly highlight the small number of studies that have looked at the effects of providing artificial roosts on bat populations, by observing changes in bat numbers over time, preferably in areas with and without bat boxes.

We would recommend a systematic review of this subject.

Key messages

Provide artificial roost structures for bats

We found 22 replicated studies of artificial roost structures from across the world. 21 studies show use of artificial roosts by bats. One study in the USA found that bats did not use the bat houses provided. 15 studies show varying occupancy rates of bats in artificial roost structures (3–100%). Two studies in Europe found an increase in bat populations using bat boxes in forest and woodland.

Eight studies looked at bat box position. Three of four studies found that box orientation and exposure to sunlight are important for occupancy. Two studies found more bats occupying bat boxes on buildings than trees. Two studies found more bats occupying bat boxes in farm forestry or pine stands than in native or deciduous forest. 11 studies looked at bat box design, including size, number of compartments and temperature, and found varying results.

11.1 Provide artificial roost structures for bats

- We found 22 replicated studies of artificial roost structures in Europe, North and South America, and Australia. In 21 of these studies[1–3, 5–22] artificial roosts were used by bats. In one study in the USA[4] bats displaced from a building did not use any of 43 bat houses of four different designs.

- Fifteen studies provide occupancy rates of artificial roosts by bats with varied results. Seven studies[2, 9, 11, 13, 14, 19, 21] found bats occupying less than half of bat boxes provided (4–49%). One study in Spain[3] showed low occupancy of bird boxes by bats (3%). Six studies[6, 8, 10, 12, 15, 16] showed bats occupying more than half of bat boxes or artificial roosts provided (57–87%). One study in Costa Rica[18] found bats occupying 100% of simulated hollow tree trunks in group sizes similar to those in natural roosts.

- One study in the UK[1] found that a bat population using bat boxes in a woodland doubled over ten years. One study in Poland[19] found that the number of bats using bat boxes in a forest increased more than eightfold over three years.
- Eight studies looked at the position of bat boxes. Three of four studies in Europe and the USA found that orientation and/or the amount of exposure to sunlight were important for bat occupancy[7, 9, 15, 16]. One study found an effect of bat box height that varied between species[2], and one found no effect of height[9]. Two studies found higher occupancy of bat boxes on buildings than on trees[13, 15]. One Australian study[11] found that bat boxes were occupied more often in farm forestry sites than in native forest, and a study in Poland[14] found higher occupancy in pine relative to mixed deciduous stands.
- Eleven studies in Europe and the USA looked at bat box design. One of two studies in Spain and the USA found higher occupancy rates in larger bat boxes[9, 16]. One study in the USA found that bats used both resin and wood cylindrical artificial roosts[10]. One study in the UK found higher occupancy rates in concrete than wooden bat boxes[2]. In a study in Spain more bats occupied bat boxes that had two compartments than one compartment[15]. One study in the USA found that four of nine bat box designs were occupied by bats[17]. One study in the UK found bats selecting three of five bat box types[22]. One study in east Lithuania found that bat breeding colonies occupied standard and four/five chamber bat boxes and individuals occupied flat bat boxes[20]. Three studies found that warmer bat boxes had higher occupancy rates or were used by more bats than cooler boxes[7, 8, 12].
- Four studies[9, 11, 21, 22] found that up to 37% of bat boxes were used by birds, marsupials or invertebrates.

A replicated study in 1975–1985 in a mature coniferous forest in Suffolk, UK (1) found that the total population of brown long-eared bats *Plecotus auritus* (males, females and juveniles) occupying bat boxes doubled over the study period from 73 to 140 bats. A total of 480 bat boxes were installed, but the proportion of boxes occupied by bats is not given. Bats roosted in the boxes both individually and in clusters of up to 20 bats. Bat boxes (10 cm x 15 cm x 15 cm internal dimensions) were constructed from untreated wood and installed in 1975 on 60 evenly spaced trees and arranged into two groups of four boxes (each facing north, south, east and west) on each tree at 3 or 5 m high. In 1984 and 1985, boxes were redistributed across ten new sites within the forest. Boxes were checked and bats removed for identification and ringing two to four times every year from 1976 to 1987. The authors note that the number of bats present will be underestimated by the methods used in the study.

A replicated site comparison study from 1985 to 2005 at 52 woodland sites in the UK (2) found an overall bat box occupation rate of 8.7% (5,986 boxes occupied of 68,715 box inspections). Occupancy rates were higher in bat boxes in the west of the UK (15% in Devon and Wales) than in the east (4% in the Midlands and eastern England). Occupancy rates were higher in summer (10% in August and September) than winter (2% in February). Concrete bat boxes had higher occupancy rates than wooden boxes, with types 1ff and 2fn being occupied the most (90% of records). Occupancy rates, bat counts and species counts were higher in bat boxes established for more than 4 years (18% occupancy, 60 bats/100 box inspections, 15 species/100 box inspections) than boxes established for less than one year (8% occupancy, 22 bats/100 box inspections, 6 species/100 box inspections). Occupancy rates were higher for Natterer's bats *Myotis nattereri* in lower bat boxes (3% at ≤ 4 m, 1.6% at ≥ 7 m), and higher for

11.1 Provide artificial roost structures for bats   72

noctule bats *Nyctalus noctula* in higher bat boxes (5% at ≤ 4 m, 7.2% at ≥ 7 m). Bat boxes were installed on a total of 1,410 trees across 52 sites (with 10–208 trees/site). Ten different types of bat box were included in the study but were not installed systematically (1ff, 1fs, 1fw, 2f, 2fn, SW, Wedge, Martin, CJM and Messenger). Most trees had two bat boxes installed (120 trees had 1 bat box, and 23 had at least 5). A total of 3,024 boxes were inspected and 68,715 inspections were made. Inspections were made in one month intervals but not all boxes were inspected monthly or yearly. Due to an unbalanced design, subsets of the data were used for analysis.

A replicated study between May and November 1989 in a 130 ha coniferous forest plantation in Spain (3) found brown long-eared bats *Plecotus auritus* using bird boxes as day roosts. A total of 197 bats (49 males, 75 females and 73 juveniles) were found in 3% of bird box checks and bat droppings in 8% of checks. One hundred and twenty one bats were found roosting individually in the boxes (53% were males) and 31 groups of bats of 2–13 individuals were found (94% were females or juveniles). Bird boxes (dimensions not given) were installed in rows 50–70 m apart with an average density of 4 boxes/hectare in 1988 (total number of boxes not given). Between March and November 1989, 5,674 box checks were made with 2–12 days/visit. Bats were removed from boxes for identification, collection of biometric data and to be ringed.

A replicated study in May to August 1988–1990 at a large urban institute in New York, USA (4) found that displaced little brown myotis *Myotis lucifugus* did not use any of 43 bat houses of 4 different designs and sizes. The 4 designs tested were 20 very small bat houses (longest dimension < 0.4 m, volume 0.002 m^2), 8 small bat houses (20 cm x 15 cm x 15 cm with partitioned spaces), 11 Bat Conservation International (BCI) style bat houses (50 cm x 20 cm x 15 cm) and 4 large 'Missouri' style bat houses (2.3 m x 1 m x 1 m with partitioned spaces below and an attic-like space above). All bat houses were placed facing different directions. The very small bat houses were placed 3–4 m above the ground on trees, the small and BCI style bat houses were placed 2–7 m above the ground on the walls of buildings and the 'Missouri' style bat houses were placed on building roofs.

A replicated study from 1977 to 1993 in a 360 km^2 area of mixed woodland near Wareham, UK (5) found a total of 1,662 bats of 3 species occupying up to 500 bat boxes (occupancy rates not given) at 20 sites (976 brown long-eared bats *Plecotus auritus*, 355 common pipistrelles *Pipistrellus pipistrellus*, and 286 Natterer's bats *Myotis nattereri*). Since 1976, approximately 500 timber bat boxes (10 cm x 15 cm x 15 cm internal dimensions) were installed across the area. Each site comprised of 6 trees with 3 boxes per tree, facing north, south-east and south-west, 2.5–3 m above the ground. Adjacent sites were 0.5–2.75 km apart. Boxes were checked and bats ringed approximately four times a year in March–October from 1977 to 1993.

A replicated study in June–September 1997 and 1998 in coniferous forests in Oregon, USA (6) found that bats used 13 out of 15 (87%) bat boxes installed under 15 flat bottom bridges along 5 large streams. Within a year of installation, ten boxes were used by bats. Bats were observed day roosting in 5 different boxes on 14 occasions (all solitary bats except for one group of 8 individuals). Guano was collected from traps beneath 12 different boxes on 1–16 occasions. Wooden boxes (60 cm long x 60 cm wide x 30 cm deep) with 8 boards placed inside (12 mm or 19 mm apart) to form crevices were fixed to the underside of the bridges between September 1996 and May 1997. Day roosting bats were counted with a spotlight and guano traps were checked weekly. Bridges varied in size from 230–475 m width, 11–27 m length, and 3–6 m above the water.

A replicated study in 1991–1993 in an urban area of Pennsylvania, USA (7) found that big brown bats *Eptesicus fuscus* and little brown bats *Myotis lucifugus* used pairs of bat boxes at five out of nine sites when they had been excluded from buildings. At the four sites where boxes were not used, bats either re-entered the building, found new roosts in nearby buildings or disappeared. All bat boxes that were occupied were positioned in a south-eastern or south-western aspect and received at least seven hours of direct sunlight. Unoccupied bat boxes received less than five hours of direct sunlight. More big brown bats (at one site) were found in horizontal bat boxes (minimum of 47 bats emerging/night early summer, minimum of 67 emerging/night late summer) than vertical boxes across the summer (no bats emerging early summer, minimum of 11 bats emerging/night late summer). Little brown bats (at 5 sites) were found more in horizontal boxes in early summer (average 186 minimum bats emerging/night/site from horizontal boxes, average 79 minimum bats emerging/night/site from vertical boxes) and more in vertical boxes in late summer (average 58 minimum bats emerging/night/site from horizontal boxes, average 132 minimum bats emerging/night/site from vertical boxes). Differences were not tested for statistical significance. Horizontal bat boxes had significantly higher maximum temperatures than vertical boxes in the afternoon (average 2°C warmer) and early evening (average 1°C warmer). Each site had a maternity colony of at least 30 bats that were excluded from the buildings by homeowners from 1991 to 1992. Homeowners at each site were provided with a pair of wooden bat boxes (76 cm x 30 cm x 18 cm) and instructed to install one horizontally (30 cm tall) and one vertically (76 cm tall) side by side on the building within 5 m of the primary bat entrance.

A replicated study in April–November 1996 in deciduous forest in Bavaria, Germany (8) found that 21 marked female Bechstein's bats *Myotis bechsteinii* within a colony roosted in 43 of 75 bat boxes (57% occupancy). Out of 23 pairs of black and white boxes, females roosted significantly more often during and after lactation in black bat boxes (186 'bat days' [the sum of the number of individuals found over all survey days] during lactation, 90 'bat days' after lactation) than white bat boxes (134 'bat days' during lactation, 22 'bat days' after lactation), and more in sun exposed boxes (276 'bat days' during lactation, 112 'bat days' after lactation) than shaded boxes (44 'bat days' during lactation, no 'bat days' after lactation). Before giving birth, females roosted more in shaded locations (111 'bat days') than sunny locations (43 'bat days') but did not show a significant preference for black or white boxes (76 and 78 'bat days' respectively). Boxes of each colour were significantly warmer in sunny locations (black average 22°C, white average 20°C) than boxes in the shade (black average 18°C, white average 17°C), and black bat boxes were significantly warmer than white boxes. Seventy five bat boxes (type Schwegler 2FN) were originally installed between 1987 and 1993 over a 0.4 km^2 area. In 1996, 52 boxes were hung in pairs (one painted white, the other remained black) side by side on 26 trees (half at shaded sites, half on trees exposed to the sun). All bat boxes were checked daily and roost temperatures taken.

A replicated study in March to October 1996–1998 in a 60 ha pine grove *Pinus sylvestris* in Guadalajara, central Spain (9) found bats occupying 8% of boxes and bat droppings in 2% of boxes checked. Larger bat boxes were occupied significantly more (9%) than smaller bat boxes (7%). The height and orientation of boxes did not significantly affect bat occupation. A quarter of bat boxes were parasitized with invertebrates (arachnids and hymenoptera), had nesting birds in them, or were damaged by arboreal birds, with a higher proportion of large bat boxes affected (28%, small boxes 16%). The larger boxes were based on the 'Richter II' model (external dimensions: 40 cm height x 25 cm length x 22 cm width, internal capacity: 3,600 cm^3). The smaller boxes were based on the 'Stratmann FS 1' model (external dimensions: 40 cm height x 30 cm length x 11 cm width, internal capacity: 2,000 cm^3). During April

1996, 203 bat boxes were installed on trees (108 large, 95 small) at heights of 2.9–5.5 m in rows spaced 50 m apart with an average density of 4 boxes/ha. Sixteen surveys with 2,134 total box visits were carried out from August to October in 1996 and March to October 1997 and 1998. Bats captured in the boxes were brown long-eared *Plecotus auritus* (176 out of 178) and two common pipistrelles *Pipistrellus pipistrellus*. Recordings made with bat detectors in the study area showed the presence of Leisler's bats *Nyctalus leisleri*, Daubenton's bats *Myotis daubentonii*, Natterers' bats *Myotis nattereri* and serotine bats *Eptesicus serotinus*.

A replicated, controlled study in 1999–2000 in Fort Valley Experimental Forest, Arizona, USA (10) found that bats used 17 of 20 artificial roosts (8 resin and 9 wood) placed on snags in thinned (10 roosts) and unthinned (8 roosts) pine stands. Bats did not roost more often in natural control snags (five roosts). There was no difference in the use of two artificial bat roost designs (resin and wood, both 60 x 60 cm cylindrical designs). Resin roosts were made from polyester moulds shaped and painted to resemble exfoliating bark. Wood roosts were made from treated hardboard. Five resin and five wood artificial roosts were placed 2–4 m above the ground on snags in 3 unharvested stands and 3 thinned stands with a natural control roost on a snag at least 75 m away from each artificial roost. Nets below roosts were checked for guano 3–4 times from July to August in 1999 and 2000. Bats were observed on two occasions: two big brown bats *Eptesicus fuscus* in a wood roost, and a maternity colony of at least seven long-eared myotis *Myotis evotis* in a resin roost.

A replicated study in 1996–2000 in three farm forest plantations and one native forest in Queensland, Australia (11) found that 19 of 96 bat boxes (20%) were used by Gould's long-eared bats *Nyctophilus gouldi* as maternity and other roosts. More bat boxes were occupied at two farm forestry sites in more fragmented landscapes than in native forest (no boxes used) and one of the farm forestry sites bordering it (one box used). Approximately 20 other bat species were known to occur in the study area but did not use the bat boxes. Four marsupials occupied 30% of bat boxes: feathertail gliders *Acrobates pygmaeus* (16 boxes), sugar gliders *Petaurus breviceps* (10 boxes), squirrel gliders *Petaurus norfolcensis* (4 boxes) and the yellow-footed marsupial mouse *Antechinus flavipes* (1 box). Bat boxes were made from laminated plywood built to the British Tanglewood Wedge design (40 cm long x 20 cm wide x maximum of 18.5 cm deep). Twenty four boxes were attached to trees at each site 3 m or 6 m above the ground, evenly spaced and in different aspects. Boxes at each site were checked five to nine times between April 1996 and November 2000.

A small replicated study in May–June 2001 in Alentejo and Algarve, Portugal (12) found that soprano pipistrelles *Pipistrellus pygmaeus* used six of nine bat boxes present at three urban sites. More bats were seen emerging from black bat boxes (maximum of 38) than grey boxes (maximum of 6), although this difference was not statistically tested. No bats were seen to emerge from white bat boxes. The internal temperatures of different coloured bat boxes varied significantly (average maximum temperatures: black 37°C, grey 34°C and white 28°C). Maximum daily temperatures inside black bat boxes did not differ significantly to those in roosts in the attics of nearby buildings. Three bat boxes (painted black, grey or white, all three compartment Bat Conservation International models) were placed facing south side by side at each site 20 m from maternity roosts. Bat box temperatures were monitored using sensors and data loggers. Bat boxes were checked and emerging bats counted weekly.

A replicated study in May–September 1997 in Colorado, USA (13) found bats occupying 11 out of 95 bat houses (12% occupancy rate) at multiple sites. Big brown bats *Eptesicus fuscus* occupied six boxes *Myotis* spp. two boxes and little brown bats *Myotis lucifugus* one box. All bat houses were occupied by 1 or 2 individuals, except 1 colony of 20 big brown bats. In areas where bats roosted prior to bat house installation, occupancy rate increased to 64%. The likelihood of bat house occupation increased when bat houses had large landing areas, were mounted on buildings rather than trees, and in areas of low canopy cover and human disturbance. No bat houses mounted on trees were occupied. Bat houses were installed in preserved areas (47), remote campgrounds (8), rural farmland (39) and irrigated farmland (1) placed on trees (40), buildings (42) and poles (13). Details of the locations of occupied bat houses are not given. Bat houses used were different sizes, colours and designs. Bat houses were checked for occupancy and guano on the ground below at 15 or 30 day intervals.

A replicated study in 1998–2001 in 3 different forest stands (pine, beech and oak-beech) in a mixed forest in Poland (14) found that an average of 4 of 102 bat boxes (4%) were occupied by bats during each box check (Nathusius' pipistrelles *Pipistrellus nathusii* or brown long-eared bats *Plecotus auritus* roosting individually or in groups). The number of boxes occupied and the number of Nathusius' pipistrelles occupying bat boxes was significantly higher in the pine stand (maximum 9 boxes occupied, 36 bats/100 boxes) than the two deciduous stands (maximum 2 boxes and 1 bat/100 boxes in beech, 1 box and 0.2 bats/100 boxes in oak-beech). Bat boxes were occupied within 2 months in the pine stand, but more slowly in the beech and oak-beech stands (13 months or more). In 1998, 34 wooden bat boxes (Stratmann design, 40 cm x 13 cm x 4 cm) were installed per stand. Bat boxes were checked every ten days in July–September 1998–1999, every two weeks in April–June 1999 and for two days in August 2001. The pine forest plot was checked additionally twice in July–August 2000. Birds nested in one bat box, woodpeckers destroyed 3 boxes, and wasp nests were found in 12 boxes.

A replicated study in 1999–2004 in a wetland on an island in Catalonia, Spain (15) found that soprano pipistrelles *Pipistrellus pygmaeus* used 69 bat boxes of two different designs with an average occupancy rate of 71%. During at least one of the four breeding seasons recorded, 96% of boxes were occupied and occupation rates by females with pups increased from 15% in 2000 to 53% in 2003. Bat box preferences were detected in the breeding season only, with higher abundance in east-facing bat boxes (average 22 bats/box vs. 12 bats/box west-facing), boxes with double compartments (average 25 bats/box vs. 12 bats/box single compartment) and boxes placed on posts (average 18 bats/box) and houses (average 12 bats/box). Abundance was low in bat boxes on trees (average 2 bats/box). A total of 69 wooden bat boxes (10 cm deep x 19 cm wide x 20 cm high) of 2 types (44 single and 25 double compartment) were placed on 3 supports (10 trees, 29 buildings and 30 electricity posts) facing east and west. From July 2000 to February 2004, the boxes were checked on 16 occasions. Bats were counted in boxes or upon emergence when numbers were too numerous to count within the box.

A replicated study in 1997–2004 in 66 rural agricultural areas in California, USA (16) found that bats used 141 out of 186 available bat houses, with an overall occupancy rate of 76% (48% by groups and 28% by individuals). Five bat species were recorded, with the Brazilian free-tailed bat *Tadarida brasiliensis* and *Myotis* spp. accounting for the majority of bat house occupancy (67% and 26% respectively). Size, colour and height of the bat houses did not affect bat occupancy. Bat colonies (average of 64 bats) were more likely to use bat houses that were shaded or exposed to the morning sun, mounted on structures such as houses and that were within a quarter of a mile of a water source. Individual bats were more likely to use bat houses that were mounted on poles and exposed to the full or afternoon sun. Bat houses were not likely to become occupied if colonies had not moved in within the first two years. All bat houses were plywood with one or more chambers and categorized as small (< 90 cm roosting space) or large (> 90 cm roosting space). Bat houses were mounted singly, side by side or back to back on barns, sheds, poles, bridges or silos. All bat houses were placed within 4 km of a water source and 2–9.5 m high. Houses were placed in different orientations and painted light, medium and dark colours. Bat houses were inspected yearly, with occurrence, number and bat species recorded.

A replicated study in 1992–1999 in several small woodlots surrounded by agricultural, industrial and residential areas in Indiana, USA (17) found that 4 out of 9 artificial roost designs were used by a total of 709 bats over the 7 year study. The designs were single box (428 bats), triple box (210 bats), shake garland (96 bats) and Missouri-style bat boxes (65 bats). Five bat species used the artificial roosts both individually and in groups, with northern myotis *Myotis septentrionalis* using them most frequently (690 out of the total 709 bats). From 1992 to 1994, 3,204 artificial roosts of nine designs were installed. Single boxes (715) were 'bird house' style attached to deciduous trees. Triple boxes (259) were three single boxes surrounding deciduous trees. Single shakes (697) consisted of a pair of overlapping cedar shingles nailed to a tree. Shake garlands (842) had 10–20 shakes encircling deciduous tree trunks. Missouri style boxes (56) were 0.9 m x 1.8 m. Tarpaper boxes (30) were wooden (0.9 m x 0.9 m) and lined with tarpaper. Plastic/tarpaper skirts (176) had a length of tarpaper/plastic folded over and wrapped around a tree. Exfoliations (338) were loosened bark with the lower end wedged. Moved trees (91) were trees greater than 25 cm (diameter) at breast height which were topped and moved to loosen bark. Missouri style and tarpaper boxes were placed on posts 2.4 m high. The remaining structures were placed 3–11 m up in trees. The majority of the artificial roosts were in shaded areas. All structures were checked at least once a year and bats were captured and identified to species when bats were present.

A replicated, controlled study in 2000–2006 in an area of tropical wet forest and pasture in the Caribbean lowlands, Costa Rica (18) found that bats colonized all 45 artificial roosts installed in 2 different habitats (22 in a continuous forest habitat and 23 in small fragments or tree stands in agricultural habitat). Average colonization time was three weeks in both continuous forest and disturbed habitat. Ten bat species occupied the artificial roosts. Five nectar- or fruit-eating bat species colonized the artificial roosts permanently in group sizes similar to those in natural roosts (Pallas' long-tongued bat *Glossophaga soricina*, Commissaris's long-tongued bat *Glossophaga commissarisi*, Seba's short-tailed bat *Carollia perspicillata*, Sowell's short-tailed bat *Carollia sowelli*, chestnut short-tailed bat *Carollia castanea*). Artificial roosts were simulated hollow tree trunks made from sawdust concrete slabs forming a square box and installed in the shade. Twenty-four roosts were 54 cm x 54 cm x 194 cm and 21 roosts were 74 cm x 74 cm x 154 cm. Natural roosts were found by a systematic line transect search. Roosts were frequently inspected (every 42 days on average) and bats were captured on 105 occasions using mist nets near roosts for identification.

A replicated study in 2005–2008 in a mixed forest in Poland (19) found an increase in the occupancy rate of 70 bat boxes and the number of individuals using the bat boxes (from 17 bats, 13% occupancy in 2005 to 42 bats, 49% occupancy in 2006). Four bat species were found in the boxes: greater mouse-eared bat *Myotis myotis*, common noctule *Nyctalus noctula*, Nathusius' pipistrelle *Pipistrellus nathusii* and brown long-eared bat *Plecotus auritus*. In 2007–2008, bat boxes were colonized first by brown long-eared bats in March and last occupied in October by common noctules. Nathusius' pipistrelles were the most abundant species that used bat boxes (74% of records from May to September) and were found in the largest clusters in July (14 individuals). The highest occupancy rate of all bat species (33 of 69 boxes) and number of individuals (128) was during August. In 2003, 70 wooden Stratmann bat boxes (internal dimensions 25 cm x 25 cm x 7 cm) were installed on trees up to 30 m from a forest road 2.5–3 m above the ground with a south-easterly orientation. In 2005 and 2006, boxes were checked once in August and from March 2007 to February 2008 boxes were checked monthly.

A replicated study in May–October 2009 in 13 different mixed or pine forests in east Lithuania (20) found that 6 bat species used bat boxes of 4 designs (occupancy rates are not given). The most abundant bat species that used the bat boxes were Nathusius' pipistrelle *Pipistrellus nathusii* (79% of all bats recorded and occupying boxes at all 13 sites) and soprano pipistrelle *Pipistrellus pygmaeus* (18% of all bats and occupying boxes at 7 of the 13 sites). The remaining bat species, the pond bat *Myotis dasycneme*, brown long-eared bat *Plecotus auritus*, common noctule *Nyctalus noctula* and northern bat *Eptesicus nilssonii*, accounted for 2% of bats using the bat boxes. Breeding colonies of Nathusius' pipistrelles and soprano pipistrelles were found in standard and four/five chamber bat boxes. Flat bat boxes were not used by breeding colonies, but were the only type of bat box in which all six species were found. In total, 504 bat boxes (30–60 installed in each area) were tested: 250 standard boxes (25 cm x 15 cm x 10 cm), 168 flat boxes (35 cm x 4 cm x 15 cm), 27 4 chamber (30 cm x 15 cm x 15 cm), and 59 5 chamber boxes (55 cm x 35 cm x 19.5 cm). Standard and flat wooden bat boxes were installed in 2004–2008 and four/five chamber bat boxes were installed in 2007–2008. All boxes were attached to trees facing south-east or south-west, 4–6 m above the ground and 20–200 m away from each other. Bat boxes were checked six times between May and October 2009. The number of bats present was recorded upon emergence and using bat detectors.

A replicated study in 2005–2009 in 7 sites of mixed woodland in north-east England, UK (21) found that the overall bat occupancy of bat boxes (90 in total) varied between 9% in 2006 to 18% in 2007 (12% in 2008 and 17% in 2009). The highest proportion of bat boxes occupied at one site was 27% (7 of 26 boxes). Four bat species occupied the bat boxes: *Pipistrellus* spp., brown long-eared bat *Plecotus auritus*, Natterer's bat *Myotis nattereri* and whiskered bat *Myotis mystacinus*/Brandt's bat *Myotis brandti*. In 2006, birds occupied 37% of bat boxes across the sites (most frequently blue tit *Cyanistes (Parus) caeruleus* and great tit *Parus major*). The installation of bird boxes (2–15 boxes/site) in February 2008 reduced bird occupancy of bat boxes to 17% across the sites. Woodland sites were small (< 3 ha) linear blocks with trees less than 40 years old. In 2005–2006, bat boxes (Schwegler 2FN, 16 cm diameter x 36 cm high) were installed in sets of 3 per tree, covering different aspects at least 4 m above the ground. Boxes were checked for bats in November 2006 and 2007, September 2008 and October 2009.

A replicated, controlled study in May–October 2011 and 2012 in ancient, lowland mixed deciduous woodland in Buckinghamshire, UK (22) found that brown long-eared bats *Plecotus auritus* and Natterer's bats *Myotis nattereri* favoured 3 of 5 bat box types: 1FS (33% of total occupations), 2FN (29%) and 2F (27%). The 1FF boxes were rarely used (11%), and the Apex box was not used at all. There was seasonal variation in bat occupancy rates, with a suggestion that nesting birds outcompeted bats for the 1FS boxes between May and June. Groups of Schwegler 2F, 2FN, 1FS, 1FF woodcrete boxes and 1 wooden Apex box were erected in 13 locations (5 around each tree). The box clusters were located on trees with a proven history of good box occupancy levels – part of a ten year woodland bat box scheme. The group positions were evenly spaced along a transect line of 300 m in homogenous habitat of predominantly semi-mature pendunculate oak *Quercus robur* and ash *Fraxinus excelsior* closed canopy with lapsed hazel *Corylus avellana* coppice understorey. Box temperatures were compared and found to be similar, and consistent with the ambient temperature due to the shaded nature of the sites. Aspect was experimentally controlled by progressively rotating the box positions around the tree. Over the 2 years, 156 box checks were made for each box type, with a total of 149 bat box occupations. Differences between species were discussed in the study, but were not supported by statistical analysis and sample sizes were small.

11.1 Provide artificial roost structures for bats

(1) Boyd I.L. & Stebbings R.E. (1989) Population changes of brown long-eared bats (*Plecotus auritus*) in bat boxes at Thetford Forest. *Journal of Applied Ecology*, 26, 101–112.
(2) Poulton S.M.C (2006) *An Analysis of the Usage of Bat Boxes in England, Wales and Ireland for The Vincent Wildlife Trust*. Biological and Ecological Statistical Services, Norwich, UK.
(3) Benzal J. (1990) Population dynamics of the brown long-eared bat (*Plecotus auritus*) occupying bird boxes in a pine forest plantation in central Spain. *Netherlands Journal of Zoology*, 41, 241–249.
(4) Neilson A.L. & Fenton M.B. (1994) Response of little brown myotis to exclusion and to bat houses. *Wildlife Society Bulletin*, 22, 8–14.
(5) Park K.J., Masters E. & Altringham J.D. (1998) Social structure of three sympatric bat species (Vespertilionidae). *Journal of Zoology*, 244, 379–389.
(6) Arnett E.B. & Hayes J.P. (2000) Bat use of roosting boxes installed under flat-bottom bridges in Western Oregon. *Wildlife Society Bulletin*, 28, 890–894.
(7) Brittingham M.C. & Williams L.M. (2000) Bat boxes as alternative roosts for displaced bat maternity colonies. *Wildlife Society Bulletin*, 28, 197–207.
(8) Kerth G., Weissman K. & König B. (2001) Day roost selection in female Bechstein's bats (*Myotis bechsteinii*): a field experiment to determine the influence of roost temperature. *Oecologia*, 126, 1–9.
(9) Paz O. de, Lucas J. de & Arias, J.L. (2000) Cajas refugio para quirópteros y estudio de la población del murciélago orejudo dorado (*Plecotus auritus*) en un àrea forestal de la provincia de Guadalajara. *Ecologia*, 14, 259–268.
(10) Chambers C.L., Alm. V., Siders M.S. & Rabe M.J. (2002) Use of artificial roosts by forest-dwelling bats in Northern Arizona. *Wildlife Society Bulletin*, 30, 1085–1091.
(11) Smith G.C. & Agnew G. (2002) The value of 'bat boxes' for attracting hollow-dependent fauna to farm forestry plantations in southeast Queensland. *Ecological Management and Restoration*, 3, 37–46.
(12) Lourcenço S.I. & Palmeirim J.M. (2004) Influence of temperature in roost selection by *Pipistrellus pygmaeus* (Chiroptera): relevance for the design of bat boxes. *Biological Conservation*, 119, 237–243.
(13) White E.P. (2004) Factors affecting bat house occupancy in Colorado. *The Southwestern Naturalist*, 49, 344–349.
(14) Ciechanowski M. (2005) Utilization of artificial shelters by bats (Chiroptera) in three different types of forest. *Folia Zoologica*, 54, 31–37.
(15) Flaquer C., Torre I. & Ruiz-Jarillo R. (2006) The value of bat-boxes in the conservation of *Pipistrellus pygmaeus* in wetland rice paddies. *Biological Conservation*, 128, 223–230.
(16) Long R.F., Kiser W.M. & Kiser S.B. (2006) Well-placed bat houses can attract bats to Central Valley farms. *California Agriculture*, 60, 91–94.
(17) Whitaker J.O. Jr., Sparks D.W. & Brack V. Jr. (2006) Use of artificial roost structures by bats at the Indianapolis International Airport. *Environmental Management*, 38, 28–36.
(18) Kelm D.H., Weisner K.R. & von Helversen O. (2008) Effects of artificial roosts for fruit-eating bats on seed dispersal in a neotropical forest pasture mosaic. *Conservation Biology*, 22, 733–741.
(19) Lesiński G., Skrzypiec-Nowak P., Janiak A. & Jagnieszczak Z. (2009) Phenology of bat occurrence in boxes in central Poland. *Mammalia*, 73, 33–37.
(20) Baranauskas K. (2010) Diversity and abundance of bats (Chiroptera) found in bat boxes in East Lithuania. *Acta Zoologica Lituanica*, 20, 39–44.
(21) Meddings A., Taylor S., Batty L., Green R., Knowles M. & Latham D. (2011) Managing competition between birds and bats for roost boxes in small woodlands, north-east England. *Conservation Evidence*, 8, 74–80.
(22) Dodds M. & Bilston, H. (2013) A comparison of different bat box types by bat occupancy in deciduous woodland, Buckinghamshire, UK. *Conservation Evidence*, 10, 24–28.

12 Education and awareness raising

Background
There is a universal requirement for education and awareness raising about the diversity of bats, their role in the environment and their conservation (Hutson *et al.* 2001). Education should not only be aimed at professionals but also at members of the public.

Hutson A.M., Mickleburgh S.P. & Racey P.A. (2001) *Microchiropteran bats: global status, survey and conservation action plan*. IUCN/SSC Chiroptera Specialist Group. IUCN, Gland, Switzerland and Cambridge, UK.

Key messages

Provide training to professionals
We found no evidence for the effects of providing training to professionals that come into contact with bats.

Educate homeowners about building and planning laws
We found no evidence for the effects of educating homeowners about building and planning laws in relation to bats.

Educate to improve public perception and raise awareness
We found no evidence for the effects of educating the public to improve the public perception of bats and raise awareness about bat conservation issues.

See also 'Threat: Biological resource use – Hunting – Educate local communities about bats and hunting', and 'Threat: Human disturbance – Educate the public to reduce disturbance to hibernating bats'.

12.1 Provide training to professionals

- We found no evidence for the effects of providing training to professionals who come into contact with bats.

Background
This intervention involves providing training of best practice methods to professionals who come into contact with bats such as ecologists, conservationists, tradesmen, architects and land managers. Training should be given to specific guidelines as applicable for the laws of the country and the protection status of bats.

12.2 Educate homeowners about building and planning laws

- We found no evidence for the effects of educating homeowners about building and planning laws in relation to bats.

Background
This intervention involves making homeowners aware of building and planning laws and providing them with relevant information so that they may take appropriate action when bats are found or are present in their homes. Information resources are available for homeowners in some countries, but we found no evidence as to whether members of the public are using these resources with a benefit to bats.

12.3 Educate to improve public perception and improve awareness

- We found no evidence for the effects of educating the public to improve the public perception of bats and raise awareness about bat conservation issues.

Background
Bats have long been the victims of negative public opinion due to mythology involving vampires and witchcraft, and associations with disease, such as rabies and the Ebola virus. Education programmes and events to dispel myths and to educate the public about the importance of bats and bat conservation are in place in some countries and may be benefitting bats. However, we found no studies examining the effects of education programmes on bats.

Index

abandoned mines 24
Aberdeenshire, UK 66
access points 1, 2, 3, 66
agricultural land
    agri-environment schemes 6, 12
    agroforestry 7, 13–15
    commuting routes 6, 10
    foraging habitats 6, 7–9
    organic farming 6, 10–12
    roosting sites 6, 7
agri-environment schemes 6, 12
agroforestry 7, 13–15
air traffic control sites 18
Alabama 49
Alberta, Canada 22, 42, 44
Arizona, USA 20, 74
Artibeus spp. 36, 37
Artibeus concolour (brown fruit eating bat), 36
Artibeus gnomus (gnome fruit-eating bat), 37
Artibeus lituratus (great fruit-eating bat), 36
Artibeus obscurus (dark fruit eating bat), 37
artificial hibernacula 48, 53
artificial roost structures *see* bat boxes
Australia
    agri-environment schemes 12
    Eden 50
    New South Wales 8, 45
    Queensland 74
    south western 41, 47

banana agroforestry plantations 7, 13, 14
Barbastella barbastellus
    in selectively logged forest 39
    in shelterwood harvested woodland 40
    underpass use 26
bat boxes
    colour 74
    in coniferous forests 72
    in deciduous forests 73
    design 74, 75, 76, 76–77
    marsupials in 74
    in new buildings 2
    occupancy rates 74–75, 76–77
    overview 70–71
    in pine groves 73–74
    in snags 74
    temperature 73, 74, 77
    urban 72, 73
    in woodland 71–72, 77
bat bricks 2
Bavaria, Germany 26, 73
Bechstein's bat (*Myotis bechsteinii*), 26
big brown bat (*Eptesicus fuscus*)
    artificial roost structures 73
    in brownfield sites 3
    in burned forest stands 55, 57
    and invasive plant species 59
    in shelterwood harvested woodland 40
    in thinned forest stands 43
    wind turbine fatalities 19
bird boxes 70, 72, 77
Boiga irregularis (brown tree snake) 59
Brandt's bat (*Myotis brandti*), 77
Brazil 13, 32, 35, 36, 37, 68, 75
bridges
    gantries/bat bridges 25, 28–29
    green 25, 29
    old 1, 3
Bristol, UK 11
British Columbia, Canada 45
brown fruit eating bat (*Artibeus concolour*), 36
brown long-eared bat (*Plecotus auritus*)
    artificial roost structures, 71, 72, 74, 75, 76, 77
    bat gantry use, 29
    overpass use, 28
    underpass use, 27
brown tree snake (*Boiga irregularis*) 59
brownfield sites 3
Buckinghamshire, UK 77
building works, timing of 1, 2
buildings
    artificial roost structures in 2, 72, 73

educating homeowners   79–80
roosting sites in   1, 2–3, 6, 7
timber treatments   64, 67–69
burning, prescribed   54–57

cacao agroforestry plantations   7, 13–14
California, USA   20, 75
Canada
   agri-environment schemes   12
   Alberta   22, 42, 44
   British Columbia   45
   Ontario   50
   Rocky Mountains   52
Carollia spp.   36, 37, 38, 76
Carollia brevicauda (silky short-tailed bat), 36, 37
Carollia castanea (chestnut short-tailed bat), 76
Carollia perspicillata (Seba's short-tailed bat), 36, 38, 76
Carollia sowelli (Sowell's short-tailed bat),   76
Cascade mountains, USA   41
caves
   fungal infections   59, 61–62
   gates to restrict public access   48–52
   legal protection   49, 53
   microclimate   49, 52
   public education   49, 53
   restrictions on recreational visitors   49, 52–53
chestnut short-tailed bat (*Carollia castanea*), 76
Chicago   4, 55, 59
chlorinated hydrocarbons   65, 67
coffee agroforestry plantations   7, 13, 14–15
Coleura seychellensis (Seychelles sheath-tailed bat)   59
Colorado, USA   51, 74
Commissaris's long-tongued bat (*Glossophaga commissarisi*),   76
common pipistrelle (*Pipistrellus pipistrellus*)
   activity over green roofs,   5
   artificial roost structures,   72, 74
   on organic farms,   11
   radar as deterrent on wind turbines,   19
   timber treatment safety,   68, 69
   use of road crossing structures,   27, 28, 29

commuting routes
   agricultural   6, 10
   development and   1, 5
   light pollution   63, 66–67
controlled burning   54–57
Costa Rica   7, 13, 14, 70, 76
culling   61
Cumbria, UK   27

dark fruit eating bat (*Artibeus obscurus*),   37
Daubenton's bat (*Myotis daubentonii*),   19, 49, 74
Denver, USA   3
development, residential and commercial   2–5
Devon, UK   71
Douglas fir *(Pseudotsuga menziesii)*   42

East Lithuania   77
eastern horseshoe bat (*Rhinolophus megaphyllus*),   50
eastern red bat (*Lasiurus borealis*)
   in burned forest stands   55, 57
   and invasive plant species   59
   responses to woodland restoration   4
   in shelterwood harvested woodland   40
   in thinned forest stands   43
   wind turbine fatalities   23
echolocation   19
Eden, Australia   50
edge habitats   43
education *see* public education
effluent treatments   63, 64–65
England, UK   4, 10, 66, 71, 77
Eptesicus fuscus (big brown bat)
   artificial roost structures   73, 74
   in brownfield sites   3
   in burned forest stands   55, 57
   and invasive plant species   59
   in shelterwood harvested woodland   40
   in thinned forest stands   43
   wind turbine fatalities   19
Eucalyptus forests   8, 41, 47

farming *see* agricultural land
feral cats   59
fertilizers   10, 63, 65
fire suppression   4, 55, 59

Index

Florida, USA 49, 57
foraging areas
    invasive plant species 58–59
    light and noise pollution 65
    protecting or creating wetlands as 6, 9
    provision in urban areas 1, 4–5
    retained or planted trees 6, 7–9
    sewage treatment works 64
forest bat species 13, 15, 35, 39, 42, 47
forestry *see also* trees
    forested corridors/buffers 33, 45–47
    managing forest or woodland edges for bats 33, 43
    prescribed burning 54–57
    replanting native trees 33, 44
    residual tree patches 33, 44–45
    retaining deadwood/snags 33, 43–44
    selective/reduced impact logging 32, 35–39
    shelterwood cutting 32, 39–41
    thinning 33, 41–43
fruit-eating bat species 13, 14, 36, 37–38, 76
fungal infections 60–62
fungicides 64, 67, 68

gantries 25, 28–29
Geomyces destructans see Pseudogymnoascus destructans
Germany 9, 26, 59, 73
Giswil, Switzerland 30
Glossophaga spp. 36, 37, 76
Glossophaga commissarisi (Commissaris's long-tongued bat), 76
Glossophaga soricina (Pallus's long-tongued bat), 36, 37, 76
gnome fruit-eating bat (*Artibeus gnomus*), 37
Gould's long-eared bat (*Nyctophilus gouldi*) 41, 47, 74
gray myotis (*Myotis grisescens*), 49, 50
great fruit-eating bat (*Artibeus lituratus*), 36
greater horseshoe bat (*Rhinolophus ferrumequinum*), 60
greater mouse-eared bat (*Myotis myotis*), 76
Greece 10, 11
green bridges 25, 29
green roofs 4, 5
guano harvesting 32, 34–35

heliponds 9
Heller's broad-nosed bat (*Platyrrhinus helleri*) 37
hibernation sites *see also* roosting sites
    abandoned mines 24
    artificial hibernacula 48, 53
    cave gates to restrict public access 48–52
    guano harvesting 34
    legal protection 24, 48, 53
    microclimate 48, 52
    public education 48, 53
hoary bat (*Lasiurus cinereus*) 22, 23
homeowners, education for 79–80
homing behaviour 60
hop-overs 25, 30
hunting 32, 33–34

Indiana bat (*Myotis sodalis*), 56
Indiana, USA 49
insecticides 14, 63, 65, 67, 68
insectivorous bat species 13, 14, 15, 33
insects
    above sewage treatment works 64, 65
    on brownfield sites 3
    prescribed burning and 56, 57
    UV lights and 67
invasive plant species 58–59
invasive predators 58, 59
Ireland 26, 27, 28

Kentucky, USA 56

Landau, Germany 9
Lasionycteris noctivagans (silver-haired bat)
    in burned forest stands 55
    and invasive plant species 59
    in residual tree patches 45
    responses to woodland restoration 4
    in shelterwood harvested woodland 40
    in thinned forest stands 42
    wind turbine fatalities 22, 23
Lasiurus borealis (eastern red bat)
    in burned forest stands 55, 57
    and invasive plant species 59
    responses to woodland restoration 4
    in shelterwood harvested woodland 40
    in thinned forest stands 43

wind turbine fatalities 23
Lasiurus cinereus (hoary bat) 22, 23
leaf-nosed bat species 13, 14
legislation
    to control chemical effluents 63, 65
    guano harvesting 32, 35
    hunting 32, 34
    protection of hibernation sites 24, 48, 53
Leisler's bat (*Nyctalus leisleri*), 26, 27, 28, 74
lesser horseshoe bat (*Rhinolophus hipposideros*), 27, 30, 60, 66
lesser short-tailed bat (*Mystacina tuberculata*) 60
light pollution 63, 65–67
lighting
    around roads 26, 31
    dimmed/directional 63, 66–67
    timing 63, 67
    UV levels 63, 67
    wind turbines 16, 21
lindane 65, 67–68
little brown bat (*Myotis lucifugus*), 42, 52, 72, 73, 74
logging *see* forestry
London, UK 5
long-eared myotis (*Myotis evotis*), 74
long-legged myotis (*Myotis volans*), 52

mammal safe timber treatments 64, 68–69
marsupials 74
Maryland, USA 19
Mexico 4, 7, 13, 14, 15, 59
migratory bat species 17
mining 24
Miniopterus schreibersii (Schreiber's bat) 50
Missouri, USA 55
Myotis spp.
    activity over retention ponds 9
    artificial roost structures 71, 72, 73, 74, 75, 76, 77
    in burned forest stands 55, 56, 57
    in caves 49–51, 52
    and invasive plant species 59
    and light pollution 66–67
    on organic farms 11
    radar as deterrent on wind turbines 19
    in residual tree patches 45
    responses to woodland restoration 4
    in shelterwood harvested woodland 40
    in thinned forest stands 41, 42
    use of road crossing structures 26, 27, 28, 29
Myotis austroriparius (southeastern myotis), 50, 57
Myotis bechsteinii (Bechstein's bat), 26
Myotis brandti (Brandt's bat), 77
Myotis dasycneme (pond bat), 49, 77
Myotis daubentonii (Daubenton's bat), 19, 49, 74
Myotis evotis (long-eared myotis), 74
Myotis grisescens (gray myotis), 49, 50
Myotis lucifugus (little brown bat), 42, 52, 72, 73, 74
Myotis myotis (greater mouse-eared bat), 76
Myotis mystacinus (whiskered bat), 77
Myotis nattereri (Natterer's bat), 26, 27, 51, 71, 74, 77
Myotis septentrionalis (northern long-eared bat), 42, 52, 56, 76
Myotis sodalis (Indiana bat), 56
Myotis volans (long-legged myotis), 52
Mystacina tuberculata (lesser short-tailed bat) 60

Nathusius' pipistrelle (*Pipistrellus nathusii*) 5, 75, 76, 77
Natterer's bat (*Myotis nattereri*) 26, 27, 51, 71, 74, 77
navigation 19
nectarivorous bat species 14
New South Wales, Australia 8, 45
New York, USA 20
New Zealand 60
noctule bat (*Nyctalus noctula*) 5, 72, 76, 77
noise pollution 64, 67
northern long-eared bat (*Myotis septentrionalis*) 42, 52, 56, 76
North Carolina, USA 9
Nyctalus spp.
    activity over green roofs 5
    artificial roost structures 72, 74, 76, 77
    and light pollution 67
    on organic farms 11
    use of road crossing structures 26, 27, 28

Index

Nyctalus leisleri (Leisler's bat) 26, 27, 28, 74
Nyctalus noctula (noctule bat) 5, 72, 76, 77
Nyctophilus gouldi (Gould's long-eared bat) 41, 47, 74

Ohio, USA 40
Oklahoma, USA 50
olive groves 11
Ontario, Canada 50
Oregon, USA 20, 42, 72
organic farming 6, 10–12

Pallas' long-tongued bat *(Glossophaga soricina)* 36, 37, 76
Parnell's mustached bat *(Pteronotus parnellii)* 38
Pennsylvania, USA 20, 22, 73
pentachlorophenol (PCP) 65, 67, 68
Perimyotis subflavus 40, 43, 47, 55
pesticides 10, 63, 65, 69
pine groves, bat boxes in 73–74
Pinus spp. 46, 47, 73
Pinus sylvestris 73
Pinus taeda 46, 47
Pipistrellus nathusii (Nathusius' pipistrelle) 5, 75, 76, 77
Pipistrellus pipistrellus (common pipistrelle)
    activity over green roofs 5
    artificial roost structures 72, 74
    on organic farms 11
    radar as deterrent on wind turbines 19
    timber treatment safety 68, 69
    use of road crossing structures 27, 28, 29
Pipistrellus pygmaeus (soprano pipistrelle)
    activity over green roofs 5
    artificial roost structures 74, 75, 77
    and light pollution 66
    radar as deterrent on wind turbines 19
    use of road crossing structures 27, 28, 29
planning laws 79–80
plant species, invasive 58–59
plantain monocultures 14
Platyrrhinus helleri (Heller's broad-nosed bat) 37
Plecotus auritus (brown long-eared bat)
    artificial roost structures 71, 72, 74, 75, 76, 77
    bat gantry use 29
    overpass use 28
    underpass use 27
Poland 75, 76
pollution
    effluent treatments 63, 64–65
    light 63, 65–67
    noise 64, 67
    timber treatments 64, 67–69
Portugal 74
predators 58, 59
prescribed burning 54–57
professionals, training for 79
Pseudogymnoascus destructans 60–62
Pseudotsuga menziesii (Douglas fir) 42
Pteronotus parnellii (Parnell's mustached bat) 38
public education
    around hunting 32, 34
    hibernation sites 48, 53
    overview 79–80
Plecotus townsendii (Townsend's big-eared bat) 49
pond bat *(Myotis dasycneme)* 49, 77
pyrethroid insecticides 64, 68, 69

Queensland, Australia 74

radar deterrents (wind turbines) 16, 18–19
rainforest 13–15
rats 59
reduced impact logging 35
residual tree patches 44–45
retention ponds 9
Rhinolophus spp.
    in caves 50
    and light pollution 66
    on organic farms 11
    translocation 60
    use of road crossing structures 27, 30
Rhinolophus ferrumequinum (greater horseshoe bat) 60
Rhinolophus hipposideros (lesser horseshoe bat) 27, 30, 60, 66
Rhinolophus megaphyllus (eastern horseshoe bat) 50

rivers, pollution 63, 64
road crossing structures
    bridges 25, 28–29
    diverting to safe crossing points 26, 30–31
    habitat improvements around roads 26, 31
    lighting deterrents 26, 31
    overpasses 25, 28
    underpasses 25, 26–28
Rocky Mountains, Canada 52
roof spaces
    conservation 2
    mammal safe timber treatments 64, 68–69
roosting sites *see also* artificial roost structures; hibernation sites
    abandoned mines 24, 52
    access points 1, 2
    within buildings 1, 2–3, 6, 7
    light pollution 63, 66
    microclimate 48, 52
    old/dead trees 6, 7

Scandinavia 23
Schreiber's bat *(Miniopterus schreibersii)* 50
Scotland, UK 11, 12, 18, 19, 64, 68, 69
Scottish Rural Stewardship Scheme 12
Seba's short-tailed bat *(Carollia perspicillata)* 36, 38, 76
selective logging 35
sewage treatment works 63, 64–65
Seychelles sheath-tailed bat *(Coleura seychellensis)* 59
shelterwood cutting 32, 39–41
silver-haired bat *(Lasionycteris noctivagans)*
    in burned forest stands 55
    and invasive plant species 59
    in residual tree patches 45
    responses to woodland restoration 4
    in shelterwood harvested woodland 40
    in thinned forest stands 42
    wind turbine fatalities 22, 23
silky short-tailed bat *(Carollia brevicauda)* 36, 37
snags 33, 43–44
snakes 59
soprano pipistrelle *(Pipistrellus pygmaeus)*

activity over green roofs 5
artificial roost structures 74, 75, 77
and light pollution 66
radar as deterrent on wind turbines 19
use of road crossing structures 27, 28, 29
South Carolina, USA 36, 42, 46, 54
southern forest bat *(Vespadelus regulus)* 41, 45, 47
southeastern myotis *(Myotis austroriparius)* 50, 57
Sowell's short-tailed bat *(Carollia sowelli)* 76
Spain 71, 72, 73, 75
Suffolk, UK 68, 71
Switzerland 30, 60

tent-making bat *(Uroderma bilobatum)* 38
thinning 33, 41–43
timber treatments 64, 67–69
Townsend's big-eared bat *(Plecotus townsendii)* 49
traffic noise 65
translocation 58, 60
trees *see also* forestry
    old/dead 6, 7, 33, 43–44
    replanting natives 33, 44
    retained or planted 6, 7–9
Trinidad 37, 38
Turkey 51, 53

ultrasound deterrents (wind turbines) 16, 19–21
underground war bunkers 49
underpasses 25, 26–28
United Kingdom
    Aberdeenshire 66
    Buckinghamshire 77
    Cumbria 27
    Devon 71
    London 5
    northern England 29
    Suffolk 68, 71
urban areas
    bat boxes 72, 73
    brownfield sites 1, 3
    commuting routes 1, 5
    providing foraging areas in 1, 4–5
    roosting sites 1, 2, 3

Uroderma bilobatum (tent-making bat) 38
USA
    Arizona 20, 74
    California 20, 75
    Cascade mountains 41
    Chicago 4, 55, 59
    Colorado 51, 74
    Denver 3
    Florida 50, 57
    Indiana 49
    Kentucky 56
    Maryland 19
    Missouri 55
    New York 20, 72
    North Carolina 9
    Ohio 40
    Oklahoma 50
    Oregon 20, 42, 72
    Pennsylvania 20, 22, 73
    South Carolina 36, 42, 46, 54
    West Virginia 20, 49, 56
    wind farm fatalities 17

Vespadelus regulus (southern forest bat) 41, 45, 47
Victoria, Australia 8

Wales, UK 10, 66, 71
weather radar sites 18
West Virginia, USA 20, 49, 56
wetlands, protecting or creating as foraging habitat 6, 9
white-nose syndrome 59, 60–62
whiskered bat (*Myotis mystacinus*) 77
wind turbines
    closing off nacelles 17, 23
    design 16, 17
    distance from habitat features 16, 18
    lighting 16, 21
    placement 17, 17–18
    radar deterrents 16, 18–19
    switching off 16–17, 21–23
    ultrasound deterrents 16, 19–21
wood harvesting *see* forestry
woodland restoration, responses to 4–5

Zakynthos, Greece 11